Edited by
Artem R. Oganov

**Modern Methods of
Crystal Structure Prediction**

Related Titles

Desiraju, G. R. (ed.)

Crystal Design
Structure and Function

2003
ISBN: 978-0-470-84333-8

Westbrook, J. H., Fleischer, R. L. (eds.)

Intermetallic Compounds, Volume 1, Crystal Structures

2000
ISBN: 978-0-471-60880-6

Desiraju, G. R. (ed.)

The Crystal as a Supramolecular Entity

1996
ISBN: 978-0-471-95015-8

Edited by Artem R. Oganov

Modern Methods of Crystal Structure Prediction

WILEY-VCH Verlag GmbH & Co. KGaA

The Editor

Artem R. Oganov
Department of Geosciences and
Department of Physics and Astronomy
State University of New York
Stony Brook, NY 11794
USA
artem.oganov@sunysb.edu

All books published by **Wiley-VCH** are carefully produced. Nevertheless, authors, editors, and publisher do not warrant the information contained in these books, including this book, to be free of errors. Readers are advised to keep in mind that statements, data, illustrations, procedural details or other items may inadvertently be inaccurate.

Library of Congress Card No.: applied for

British Library Cataloguing-in-Publication Data
A catalogue record for this book is available from the British Library.

Bibliographic information published by the Deutsche Nationalbibliothek
The Deutsche Nationalbibliothek lists this publication in the Deutsche Nationalbibliografie; detailed bibliographic data are available on the Internet at <http://dnb.d-nb.de>.

© 2011 WILEY-VCH Verlag & Co. KGaA, Boschstr. 12, 69469 Weinheim, Germany

All rights reserved (including those of translation into other languages). No part of this book may be reproduced in any form – by photoprinting, microfilm, or any other means – nor transmitted or translated into a machine language without written permission from the publishers. Registered names, trademarks, etc. used in this book, even when not specifically marked as such, are not to be considered unprotected by law.

Composition Laserwords Private Ltd., Chennai, India
Printing and Binding Fabulous Printers Pte. Ltd., Singapore
Cover Design Adam Design, Weinheim

Printed in Singapore
Printed on acid-free paper

ISBN: 978-3-527-40939-6

Contents

List of Contributors IX
Introduction: Crystal Structure Prediction, a Formidable Problem XI

1   **Periodic-Graph Approaches in Crystal Structure Prediction** 1
       Vladislav A. Blatov and Davide M. Proserpio
1.1   Introduction 1
1.2   Terminology 2
1.3   The Types of Periodic Nets Important for Crystal Structure Prediction 5
1.4   The Concept of Topological Crystal Structure Representation 7
1.5   Computer Tools and Databases 10
1.6   Current Results on Nets Abundance 12
1.7   Some Properties of Nets Influencing the Crystal Structure 14
1.7.1   Symmetry of Nets and Embeddings 14
1.7.2   Relations Between Nets 17
1.7.3   Role of Geometrical and Coordination Parameters 18
1.8   Outlook 25
       References 26

2   **Energy Landscapes and Structure Prediction Using Basin-Hopping** 29
       David J. Wales
2.1   Introduction 29
2.2   Visualizing the Landscape 30
2.3   Basin-Hopping Global Optimization 36
2.4   Energy Landscapes for Crystals and Glasses 42
       References 46

3   **Random Search Methods** 55
       William W. Tipton and Richard G. Hennig
3.1   Introduction 55
3.2   History and Overview 57
3.3   Methods 58

Modern Methods of Crystal Structure Prediction. Edited by Artem R. Oganov
Copyright © 2011 WILEY-VCH Verlag GmbH & Co. KGaA, Weinheim
ISBN: 978-3-527-40939-6

3.4	Applications and Results	*61*
3.5	Summary and Conclusions	*64*
	References	*65*

4 Predicting Solid Compounds Using Simulated Annealing *67*
J. Christian Schön and Martin Jansen
4.1 Introduction *67*
4.2 Locally Ergodic Regions on the Energy Landscape of Chemical Systems *68*
4.3 Simulated Annealing and Related Stochastic Walker-Based Algorithms *71*
4.3.1 Basic Simulated Annealing *71*
4.3.2 Adjustable Features in Simulated Annealing *74*
4.3.2.1 Choice of Moveclass *74*
4.3.2.2 Temperature Schedule and Acceptance Criterion *76*
4.3.2.3 Extensions and Generalizations of Simulated Annealing *77*
4.4 Examples *79*
4.4.1 Structure Prediction *80*
4.4.1.1 Alkali Metal Halides *80*
4.4.1.2 Na_3N *81*
4.4.1.3 $Mg(BH_4)_2$ *82*
4.4.1.4 Elusive Alkali Metal Orthocarbonates Balancing $M_4(CO_4)$ and $M_2O + M_2(CO_3)$, with M = Li, Na, K, Rb, Cs *83*
4.4.1.5 Alkali Metal Sulfides M_2S (M = Li, Na, K, Rb, Cs) *83*
4.4.1.6 Boron Nitride *84*
4.4.1.7 Structure Prediction of SrO as Function of Temperature and Pressure *84*
4.4.1.8 Phase Diagrams of the Quasi-Binary Mixed Alkali Halides *86*
4.4.2 Structure Prediction Employing Structural Restrictions *87*
4.4.2.1 Complex Ions as Primary Building Units *87*
4.4.2.2 Molecular Crystals *88*
4.4.2.3 Zeolites *91*
4.4.2.4 Phase Diagrams Restricted to Prescribed Sublattices *92*
4.4.3 Structure Determination *94*
4.4.3.1 Structure Determination using Experimental Cell Information *94*
4.4.3.2 Reverse Monte Carlo Method and Pareto Optimization *94*
4.5 Evaluation and Outlook *96*
4.5.1 State-of-the-Art *96*
4.5.2 Future *97*
References *98*

5 Simulation of Structural Phase Transitions in Crystals: The Metadynamics Approach *107*
Roman Martoňák
5.1 Introduction *107*

5.2	Simulation of Structural Transformations	108
5.3	The Metadynamics-Based Algorithm	110
5.4	Practical Aspects	113
5.5	Examples of Applications	115
5.6	Conclusions and Outlook	125
	Acknowledgments	126
	References	127
6	**Global Optimization with the Minima Hopping Method**	**131**
	Stefan Goedecker	
6.1	Posing the Problem	131
6.2	The Minima Hopping Algorithm	134
6.3	Applications of the Minima Hopping Method	142
6.4	Conclusions	143
	References	144
7	**Crystal Structure Prediction Using Evolutionary Approach**	**147**
	Andriy O. Lyakhov, Artem R. Oganov, and Mario Valle	
7.1	Theory	148
7.1.1	Search Space, Population, and Fitness Function	150
7.1.2	Representation	150
7.1.3	Local Optimization and Constrains	151
7.1.4	Initialization of the First Generation	152
7.1.5	Variation Operators	155
7.1.6	Survival of the Fittest and Selection of Parents	157
7.1.7	Halting Criteria	158
7.1.8	Premature Convergence and How to Prevent It: Fingerprint Function	159
7.1.9	Improved Selection Rules and Heredity Operator	161
7.1.10	Extension to Molecular Crystals	162
7.1.11	Adaptation to Clusters	162
7.1.12	Extension to Variable Compositions: Toward Simultaneous Prediction of Stoichiometry and Structure	163
7.2	A Few Illustrations of the Method	164
7.2.1	Elements	165
7.2.1.1	Boron: Novel Phase with a Partially Ionic Character	165
7.2.1.2	Sodium: A Metal that Goes Transparent under Pressure	167
7.2.1.3	Superconducting ξ-Oxygen	170
7.2.1.4	Briefly on Some of the (Many) Interesting Carbon Structures	171
7.2.2	Compounds and Minerals	172
7.2.2.1	Insulators by Metal Alloying?	172
7.2.2.2	MgB_2: Analogy with Carbon and Loss of Superconductivity under Pressure	172
7.2.2.3	Hydrogen-Rich Hydrides under Pressure, and Their Superconductivity	173

7.2.2.4	High-Pressure Polymorphs of CaCO$_3$	175
7.3	Conclusions	176
	Acknowledgments	177
	References	177

8 Pathways of Structural Transformations in Reconstructive Phase Transitions: Insights from Transition Path Sampling Molecular Dynamics 181

Stefano Leoni and Salah Eddine Boulfelfel

8.1	Introduction	181
8.1.1	Shape of the Nuclei	182
8.2	Transition Path Sampling Molecular Dynamics	183
8.2.1	First Trajectory	183
8.2.2	Trajectory Shooting and Shifting	184
8.3	The Lesson of Sodium Chloride	186
8.3.1	Simulation Strategy	187
8.3.2	Topological Models	187
8.3.3	Combining Modeling and Molecular Dynamics Simulations	190
8.3.4	The Mechanism of the B1–B2 Phase Transition	191
8.3.5	Crossing the Line: NaBr	193
8.4	The Formation of Domains	194
8.5	Structure of the B2–B1 Interfaces	197
8.5.1	Domain Formation in RbCl	199
8.5.2	Liquid Interfaces in CaF$_2$	201
8.6	Domain Fragmentation in CdSe Under Pressure	204
8.6.1	B4–B1–B4 Transformation	206
8.6.2	Defects	209
8.6.3	The Lesson of CdSe	209
8.7	Intermediate Structures During Phase Transitions	210
8.7.1	Intermediates Along the Pressure-Induced Transformation of GaN	211
8.7.2	Polymorphism and Transformations of ZnO: Tetragonal or Hexagonal Intermediate?	214
8.8	Conclusions	217
	References	218

Appendix: First Blind Test of Inorganic Crystal Structure Prediction Methods 223

Color Plates 233

Index 245

List of Contributors

Vladislav A. Blatov
Samara State University
Chemistry Department
Ac. Pavlov St. 1
443011 Samara
Russia

Salah Eddine Boulfelfel
MPI CPfS
Noethnitzerstr. 40
01187 Dresden
Germany

Stefan Goedecker
Basel University
Department of Physics und
Astronomy
Klingelbergstrasse 82
4056 Basel
Switzerland

Richard G. Hennig
Cornell University
Department of Materials
Science and Engineering
214 Bard Hall
Ithaca, NY 14853-1501
USA

Martin Jansen
Max Planck Institute for
Solid State Research
Heisenbergstr. 1
70569 Stuttgart
Germany

Stefano Leoni
MPI CPfS
Noethnitzerstr. 40
01187 Dresden
Germany

Andriy O. Lyakhov
Stony Brook University
Department of Geosciences and
New York Center for
Computational Science
Stony Brook, NY 11794-2100
USA

Roman Martoňák
Comenius University
Department of Experimental
Physics
Mlynská dolina F2
842 48 Bratislava
Slovakia

Modern Methods of Crystal Structure Prediction. Edited by Artem R. Oganov
Copyright © 2011 WILEY-VCH Verlag GmbH & Co. KGaA, Weinheim
ISBN: 978-3-527-40939-6

Artem R. Oganov
Stony Brook University
Department of Geosciences,
Department of Physics and
Astronomy and New York Center
for Computational Science
Stony Brook, NY 11794-2100
USA

and

Moscow State University
Geology Department
119992 Moscow
Russia

Davide M. Proserpio
Università degli Studi di Milano
Dipartimento di Chimica
Strutturale e Stereochimica
Inorganica (DCSSI)
Via G. Venezian 21
20133 Milano
Italy

J. Christian Schön
Max Planck Institute for
Solid State Research
Heisenbergstr. 1
70569 Stuttgart
Germany

William W. Tipton
Cornell University
Department of Materials
Science and Engineering
214 Bard Hall
Ithaca, NY 14853-1501
USA

Mario Valle
Data Analysis and
Visualization Services
Swiss National Supercomputing
Centre (CSCS)
Cantonale Galleria 2
Manno, 6928
Switzerland

David J. Wales
University of Cambridge
Department of Chemistry
Lensfield Road
Cambridge, CB2 1EW
UK

Scott M. Woodley
University College London
Department of Chemistry
Gower Street
London WC1E 6 BT
UK

Introduction: Crystal Structure Prediction, a Formidable Problem

Artem R. Oganov

The famous 1988 editorial in *Nature* by John Maddox [1] stated:

> *"One of the continuing scandals in the physical sciences is that it remains in general impossible to predict the structure of even the simplest crystalline solids from a knowledge of their chemical composition".*

The central topic of the present volume is to review the state of the art in resolving this "scandal". Crystal structure is arguably the most important piece of information about a material, as it determines – directly or indirectly – pretty much all properties of a material. Knowing the structure, one can compute a large number of properties of a material, even before it is synthesized – hence the crucial importance of structure prediction for computational materials design. When the structure is unknown and cannot be predicted, very little can be said about the material.

Until recently, it was widely believed that crystal structures are fundamentally unpredictable [1–3] – as human behavior, or earthquakes, or long-term behavior of stock exchange. However, the situation began to change dramatically in 2003–2006, and this avalanche-like development of this important field can be called a scientific revolution that continues to this day. The aim of this book is to present some of the most important modern approaches to the formidable problem of crystal structure prediction.

What do we exactly mean by "crystal structure prediction problem"? For each chemical composition there are an infinite number of possible atomic arrangements that can, in principle, be obtained in the laboratory – these correspond to all possible local minima of the free energy. Among these, at each thermodynamic conditions (pressure, temperature, chemical potential) there are a finite number of special structures, extreme in some sense – the lowest energy (i.e. the most stable structures), the highest/lowest value of some other property (hardness, density, band gap, superconducting T_c, ...), or highest rate of nucleation (corresponding to kinetically preferred phases). Prediction of these structures is a well-defined and crucially important problem. In the simplest and most important case, by crystal structure prediction we mean finding, at given P-T conditions, the stable crystal structure knowing only the chemical formula.

Modern Methods of Crystal Structure Prediction. Edited by Artem R. Oganov
Copyright © 2011 WILEY-VCH Verlag GmbH & Co. KGaA, Weinheim
ISBN: 978-3-527-40939-6

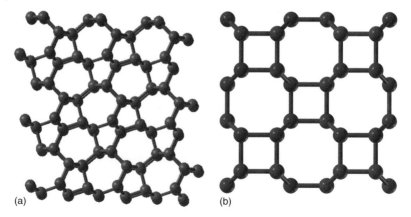

Figure 1 Structures of metastable superhard sp^3-bonded allotropes of carbon: (a) M-carbon [5] and (b) bct4-carbon [6].

Many types of approaches have been proposed to address this problem. Some are topological (as reviewed in the chapter by Blatov and Proserpio [4]) and aim at constructing the simplest topologies consistent with what we know about the chemistry of the system. This way, assuming sp^2-hybridization of carbon atoms one would arrive at 2H-graphite structure, and assuming sp^3-hybridization one would find the diamond and lonsdaleite structures – and a vast array of interesting metastable structures, including clathrates, M-carbon (Figure 1a), bct-carbon (Figure 1b) and other possible allotropes.

Topological approaches often appeal to symmetry, since in the vast majority of cases stable crystal structures do display some symmetry – the asymmetric point group 1 (or corresponding to it space group P1) is very rare (see Table 1). The ubiquity of symmetry may simplify the task of structure prediction, and not only in topological approaches.

Other approaches are based on empirical correlations and involve either *structural diagrams* [8–10] or *data mining approaches* [11, 12]. In either case, a large database of known stable crystal structures is required. While data mining approaches involve advanced machine learning concepts and are capable of predicting not only stable structures, but also the likelihood of compound formation in multinary systems (a formidable task too!), structural diagrams are much more empirical and limited in their scope. In these, one frequently uses ionic radii or the so-called "pseudopotential radii" [9], both of which (especially the latter) lack strict physical meaning and uniqueness. Instead of the "pseudopotential radii" one could use other quantities, such as the chemical scale or the Mendeleev number – the resulting empirical structure diagrams seem to have a good ability to separate structure types (e.g., Figure 2), and thus have predictive power.

The most unbiased, non-empirical and hence most generally applicable approaches are based on *computational optimization* – i.e. explicit calculations of the (free) energy and exploration of its landscape with the aim of finding the most stable arrangement of the atoms. These approaches are the main focus of

Table 1 Distribution of 280 000 chemical compounds over the 32 point groups. Note somewhat different frequencies for inorganic (I) and organic (O) compounds. (data collected by G. Johnson and published in [7]).

	I	O		I	O
1	0.67%	1.24%	422	0.40%	0.48%
$\bar{1}$	13.87	19.18	4mm	0.30	0.09
2	2.21	6.70	$\bar{4}$2m	0.82	0.34
M	1.30	1.46	4/mmm	4.53	0.69
2/m	34.63	44.81	6	0.41	0.22
222	3.56	10.13	$\bar{6}$	0.07	0.01
mm2	3.32	3.31	6/m	0.82	0.17
mmm	12.07	784	622	0.24	0.05
3	0.36	0.32	6mm	0.45	0.03
$\bar{3}$	1.21	0.58	$\bar{6}$m2	0.41	0.02
32	0.54	0.22	6/mmm	2.82	0.05
3m	0.74	0.22	23	0.44	0.09
$\bar{3}$m	3.18	0.25	m3	0.84	0.15
4	0.19	0.25	432	0.13	0.01
$\bar{4}$	0.25	0.18	$\bar{4}$3m	1.42	0.11
4/m	1.17	0.67	m3m	6.66	0.12

this book. Among the advantages are (i) the explicit calculation of the optimized quantity of interest (e.g., the energy), (ii) unbiased search techniques for exploring the energy landscape can – unlike the previously mentioned approaches, assuming knowledge of material's chemistry and likely crystal structures – arrive at completely unexpected results and truly novel structures. For instance, who would guess (based on whatever chemical knowledge) that boron under pressure would assume a NaCl-type structure composed of B_2 and B_{12} clusters with partially ionic bonding between the two? Who would guess that, when compressed to 2 million atmospheres, sodium assumes a structure unknown for any other element and becomes a transparent dielectric? Nevertheless, this is exactly what happens [13, 14], and these phenomena were first predicted using optimization techniques and only then confirmed experimentally.

When considering crystal structure prediction as an optimization problem – i.e. the problem of finding the global minimum of the energy landscape, certain properties of this landscape need to be explored. First, the number of distinct points on the landscape can be estimated as:

$$C = \left(\frac{V/\delta^3}{N}\right) \prod_i \binom{N}{n_i} \tag{1}$$

where N is the number of atoms in the unit cell of volume V, δ is a relevant discretization parameter (for instance, 1 Å) and n_i is the number of atoms of i-th

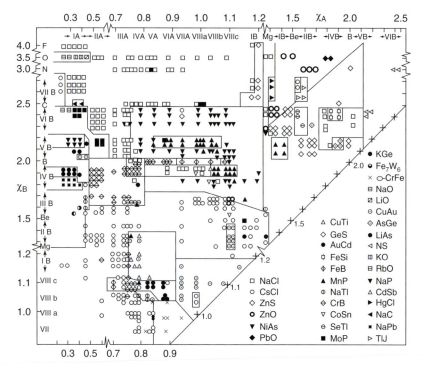

Figure 2 Pettifor's structure diagram for 574 AB compounds (from [10]).

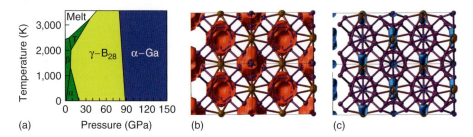

Figure 3 Boron: (a) its schematic phase diagram (from [13]) and distribution of electrons corresponding (b) bottom and (c) top of the valence band in γ-B$_{28}$ [15].

type in the unit cell. C is astronomically large (roughly, $\sim 10^N$ if one uses $\delta = 1\,\text{Å}$ and typical atomic volume of $10\,\text{Å}^3$).

It is useful to consider the dimensionality of the energy landscape:

$$d = 3N + 3 \qquad (2)$$

where $3N$-3 degrees of freedom are the atomic positions, and the remaining six dimensions are lattice parameters. For a system with 100 atoms in the unit cell, the landscape is 303-dimensional!

Equation (1) implies that the difficulty of crystal structure prediction increases exponentially with system size (or landscape dimensionality) and it thus poses an NP-hard problem (which is a shorthand of "non-deterministic polynomial-time hard", meaning that the scaling of the problem with the system size is faster than any polynomial). Such high-dimensional problems with astronomically large numbers of possible solutions imply that simple exhaustive search strategies are unfeasible.

Great simplification of the problem can be achieved if structures are relaxed, i.e. brought to the nearest local energy minima. During relaxation, certain correlations between atomic positions set in – interatomic distances adjust to reasonable values, and unfavorable interactions are avoided. The intrinsic dimensionality is thus equal to a reduced value:

$$d^* = 3N + 3 - \kappa \tag{3}$$

where κ is the (non-integer) number of correlated dimensions. Just doing relaxation, great simplifications of the global optimization problem can be achieved – for example, the dimensionality drops from 99 to 11.6 for $Mg_{16}O_{16}$ (a really simple system), while the decrease is less substantial for chemically complex systems – from 39 to 32.5 for $Mg_4N_4H_4$. To appreciate this simplification of the problem, we remind that the number of local minima depends exponentially on the intrinsic dimensionality:

$$C^* = \exp(\beta d^*) \tag{4}$$

This implies that any efficient search method must include structure relaxation (i.e. local optimization). Even simple random sampling, when combined with local optimization, can deliver correct solutions – although only for very small systems, roughly $N < 8-10$ ([17, 18], see chapter by Tipton and Hennig [19] in this volume). Much larger systems can be treated by more advanced methods, such as simulated annealing ([20, 21], see chapter by Schön and Jansen [22]), metadynamics ([23, 24], see chapter by Martoňák [25]), basin hopping ([26], see chapter by Wales in this volume [27]), minima hopping ([28], see chapter by Goedecker [29]), or evolutionary algorithms ([5, 30, 31], see chapter by Lyakhov et al. [32]). Many of the above methods rely on the fact that in usual chemical systems good (i.e. low energy) structures share some similarities, i.e. are located relatively close to each other on the landscape, forming the so-called energy funnels, low-energy regions of configuration space. This gives the landscape a benign overall shape – such as the one shown in Figure 5. Exploiting the fact (assumed for a long time by chemists, but now proven on real systems – see, e.g. [16]) that in real chemical systems there are only a few (or just one) energy funnels, allows further gains of efficiency of structure predictions. Nowadays, systems with a few hundred degrees of freedom can be treated by some of these methods – perfectly adequate for most inorganic and organic systems. Extending this limit to much larger systems may

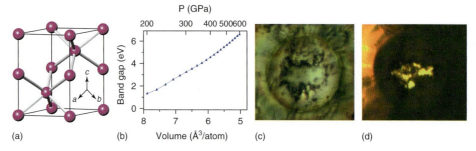

Figure 4 hP4 phase of sodium: (a) crystal structure, (b) band gap computed in the GW approximation, and optical photographs of a sodium sample at (c) 110 GPa (where sodium is a white reflecting metal) and (d) at 199 GPa (where it is a red transparent insulator). After [14].

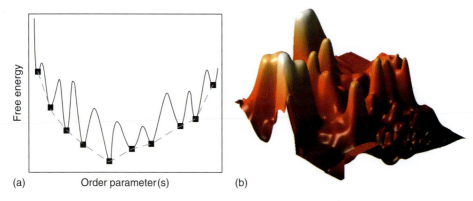

Figure 5 Energy landscape: (a) schematic illustration showing the full landscape (solid line) and reduced landscape (dashed line interpolating local minima points), (b) 2D-projection of the reduced landscape of Au_8Pd_4 (done using the method presented in [16]) showing all low-energy structures clustered in one region of configuration space.

enable us to treat biologically important systems and address such problems as protein folding. There are already some steps in this direction (see, e.g. [29]).

Next level of complexity is to ask if we can predict not just the stable structure, but also the whole set of stable chemical compositions (and the corresponding structures) in multicomponent system. This means that we are dealing with a complex landscape consisting of compositional and structural coordinates, and instead of a single ground state we should have a set of ground states located on the so-called convex hull (Figure 6). There are some encouraging steps in solving this problem [33–35].

We can also consider landscapes of properties other than the (free) energy. In this case, hybrid optimization needs to be performed – combining local optimization

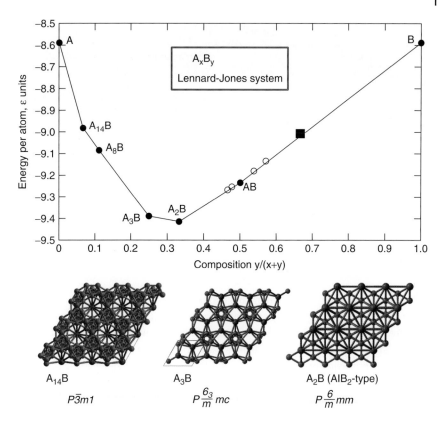

Figure 6 Examples of a simultaneous prediction of stable compositions and corresponding structures in a binary Lennard-Jones system with variable composition. Solid circles denote ground states, open circles – metastable solutions. See [35] for the potential model. From [35]. It is amazing that such a simple system gives such a wealth of complex ground states.

with respect to the energy (to find possible (quasi)equilibrium states) and global optimization with respect to the property of interest. This allows us to find structures (and compositions), corresponding to the desired values of the physical property of interest. Figure 7 gives an example of such optimization, utilizing the evolutionary algorithm [5, 30, 31] to search for the hardest possible structure of SiO_2.

This search employed the model of hardness [37] extended by Lyakhov and Oganov [36, 38], who also questioned whether diamond is the hardest carbon allotrope. The answer was that it indeed is (with the theoretical hardness of 89.4 GPa, within the error bars of to the experimental values [39]), but a number of other allotropes come close to it – for instance, lonsdaleite (89.3 GPa), bct4-carbon (theoretical hardness 84.2 GPa) and M-carbon (theoretical hardness of 83.4 GPa) shown in Figure 1. Both M-carbon [40] and bct4-carbon [6] (and private communication from a talented young researcher X.-F. Zhou, August 2009) structures were proposed as explanations for the experimentally observed new superhard allotrope

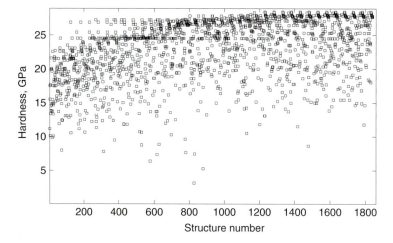

Figure 7 Evolutionary search for the hardest structure of SiO$_2$. From [36].

("superhard graphite") obtained by cold compression of graphite above 17 GPa [41]. Both structures are metastable and both match experimental observations almost equally well, but nevertheless there is a way of deciding which one is more likely to be the "superhard graphite" of Mao [41]. This brings us to the next major unsolved problem – *prediction of synthesizability of a metastable phase*. Indeed, the structure with optimal properties will frequently be metastable, and will be of interest only if it can be synthesized. This requires that the activation barrier for its formation from an available precursor (in this case, graphite) be lower than the barrier of formation of any other structure. The best approach for computing the absolute activation energies of solid-solid phase transitions is the transition path sampling

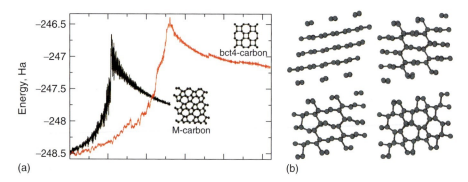

Figure 8 Synethesizability of M-carbon vs bct4-carbon: (a) energy profiles along the graphite-M-carbon (black) and graphite-bct4 (red) transition pathways and (b) the transition pathway graphite-M-carbon. The energies in given per supercell of 144 atoms. From [44]. Courtesy of S.E. Boulfelfel.

method [42], reviewed in this volume by Leoni and Boulfelfel [43]. One of the main advantages is that this method enables studies of nucleation and growth of the new phase, and its absolute activation barriers are meaningful (unlike those obtained by most other methods). As shown in Figure 8, the lower computed energy barrier clearly favors M-carbon over bct4-carbon [44].

It is my hope that this volume, reviewing most of the major methods of crystal structure prediction, all the way from topological approaches [4] to optimization methods [19, 22, 25, 27, 29, 32] and methods to appraise synthesizability of a material [43], will be useful to a wide readership of physicists, chemists, materials scientists and earth scientists. This volume also presents, in the Appendix, the first attempt to systematically compare different optimization strategies for a set of very challenging inorganic structure prediction problems [45]. The methods described in this volume should motivate further research into the structure and properties of materials, and will (probably quite soon) widely enable computational design of new functional materials. We are witnessing the dawn of a new era, where crystal structure prediction will no longer be an intractable problem.

I am grateful to Salah Eddine Boulfelfel, Andriy O. Lyakhov, Mario Valle, Feiwu Zhang, and Qiang Zhu, as well as former postdoc Yanming Ma and graduate student Colin W. Glass. I would also like to thank Wiley-VCH and its editors, in particular Anja Tschoertner, for their professionalism in preparing this book for publication. This work is supported by grants from Intel Corporation, Rosnauka (Russia, Contract No. 02.740.11.5102), Research Foundation of Stony Brook University, the National Natural Science Foundation of China (grant No. 10910263), and DARPA (grant #54751). Finally, I would like to express my gratitude to all the authors of this volume – it has been enormous pleasure to organize this book and edit it.

References

1. Maddox, J. (1988) Crystals from first principles. *Nature*, **335**, 201.
2. Gavezzotti, A. (1994) Are crystal structures predictable? *Acc. Chem. Res.*, **27**, 309–314.
3. Ball, P. (1996) Materials chemistry – scandal of crystal design... *Nature*, **381**, 648–650.
4. Blatov, V.A. and Proserpio, D.M. (2010) Periodic-Graph Approaches in Crystal Structure Prediction. **This volume**.
5. Oganov, A.R. and Glass, C.W. (2006) Crystal structure prediction using *ab initio* evolutionary techniques: principles and applications. *J. Chem. Phys.*, **124**, art. 244704.
6. Umemoto, K., Wentzcovitch, R.M., Saito, S., and Miyake, T. (2010) Body-centered tetragonal C_4: a viable sp^3 carbon allotrope. *Phys. Rev. Lett.*, **104**, 125504.
7. Newnham, R.E. *Properties of materials*. Oxford University Press (2005).
8. Mooser, E. and Pearson, W.B. (1959) On the crystal chemistry of normal valence compounds. *Acta Cryst.*, **12**, 1015–1022.
9. Burdett, J.K., Price, G.D., and Price, S.L. (1981) Factors influencing solid-state structure – an analysis using pseudopotential radii structural maps. *Phys. Rev.*, **B24**, 2903–2912.
10. Pettifor, D.G. (1984) A chemical scale for crystal structure maps. *Solid State Commun.*, **51**, 31–34.
11. Curtarolo, S., Morgan, D., Persson, K., Rodgers, J., and Ceder, G. (2003) Predicting crystal structures with data

12. Fischer, C.C., Tibbets, K.J., and Ceder, G. (2006) Predicting crystal structure by merging data mining with quantum mechanics. *Nature Materials*, **5**, 641–646.
13. Oganov, A.R., Chen, J., Gatti, C., Ma, Y.-Z., Ma, Y.-M., Glass, C.W., Liu, Z., Yu, T., Kurakevych, O.O., and Solozhenko, V.L. (2009) Ionic high-pressure form of elemental boron. *Nature*, **457**, 863–867.
14. Ma, Y., Eremets, M.I., Oganov, A.R., Xie, Y., Trojan, I., Medvedev, S., Lyakhov, A.O., Valle, M., and Prakapenka, V. (2009) Transparent dense sodium. *Nature*, **458**, 182–185.
15. Oganov, A.R. and Solozhenko, V.L. (2009) Boron: a hunt for superhard polymorphs. *J. Superhard Materials*, **31**, 285–291.
16. Oganov, A.R. and Valle, M. (2009) How to quantify energy landscapes of solids. *J. Chem. Phys.*, **130**, 104504.
17. Freeman, C.M., Newsam, J.M., Levine, S.M., and Catlow, C.R.A. (1993) Inorganic crystal structure prediction using simplified potentials and experimental unit cells – application to the polymorphs of titanium dioxide. *J. Mater. Chem.*, **3**, 531–535.
18. Schmidt, M.U. and Englert, U. (1996) Prediction of crystal structures. *J. Chem. Soc. – Dalton Trans.*, **10**, 2077–2082.
19. Tipton, W.W. and Hennig, R.G. (2010) Random search methods. **This volume**.
20. Pannetier, J., Bassasalsina, J., Rodriguez-Carvajal, J., and Caignaert, V. (1990) Prediction of crystal structures from crystal chemistry rules by simulated annealing. *Nature,*, **346**, 343–345.
21. Schön, J.C. and Jansen, M. (1996) First step towards planning of syntheses in solid-state chemistry: determination of promising structure candidates by global optimisation. *Angew. Chem. – Int. Ed.*, **35**, 1287–1304.
22. Schön, J.C. and Jansen, M. (2010) Predicting solid compounds using simulated annealing. **This volume**.
23. Martoňák, R., Laio, A., and Parrinello, M. (2003) Predicting crystal structures: The Parrinello-Rahman method revisited. *Phys. Rev. Lett.*, **90**, 075503.
24. Martoňák, R., Donadio, D., Oganov, A.R., and Parrinello, M. (2006) Crystal structure transformations in SiO_2 from classical and ab initio metadynamics. *Nature Materials*, **5**, 623–626.
25. Martoňák, R. (2010) Simulation of structural phase transitions in crystals: the metadynamics approach. **This volume**.
26. Wales, D.J. and Doye, J.P.K. (1997) Global optimization by basin-hopping and the lowest energy structures of Lennard-Jones clusters containing up to 110 atoms. *J. Phys. Chem.*, **A101**, 5111–5116.
27. Wales, D.J. (2010) Energy landscapes and structure prediction using basin-hopping. **This volume**.
28. Goedecker, S. (2004) Minima hopping: An efficient search method for the global minimum of the potential energy surface of complex molecular systems. *J. Chem. Phys.*, **120**, 9911–9917.
29. Goedecker, S. (2010) Global optimization with the minima hopping method. **This volume**.
30. Glass, C.W., Oganov, A.R., and Hansen, N. (2006) USPEX – evolutionary crystal structure prediction. *Comp. Phys. Comm.*, **175**, 713–720.
31. Lyakhov, A.O., Oganov, A.R., and Valle, M. (2010) How to predict large and complex crystal structures. *Comp. Phys. Comm.*, **181**, 1623–1632.
32. Lyakhov, A.O., Oganov, A.R., and Valle, M. (2010) Crystal structure prediction using evolutionary approach. **This volume**.
33. Wang, Y. and Oganov, A.R. (2008) Research on the evolutionary prediction of very complex crystal structures. IEEE Computational Intelligence Society Walter Karplus. Summer Research Grant 2008 Final Report. http://www.ieee-cis.org/_files/EAC_Research_2008_Report_WangYanchao.pdf
34. Trimarchi, G., Freeman, A.J., and Zunger, A. (2009) Predicting stable stoichiometries of compounds via evolutionary global space-group optimization. *Phys Rev*, **B80**, 092101.

35. Oganov, A.R., Ma, Y., Lyakhov, A.O., Valle, M., and Gatti, C. (2010) Evolutionary crystal structure prediction as a method for the discovery of minerals and materials. *Rev. Mineral. Geochem.*, **71**, 271–298.
36. Oganov, A.R. and Lyakhov, A.O. (2010) Towards the theory of hardness of materials. *J. Superhard Mater.*, **32**, 143–147.
37. Li, K.Y., Wang, X.T., Zhang, F.F., and Xue, D.F (2008) Electronegativity identification of novel superhard materials. *Phys. Rev. Lett.*, **100**, art. 235504.
38. Lyakhov, A.O. and Oganov, A.R. (2010) *Method for systematic prediction of novel superhard phases*. In prep.
39. Brookes, C.A. and Brookes, E.J. (1991) Diamond in perspective – a review of mechanical properties of natural diamond. *Diamond and Rel. Mater.*, **1**, 13–17.
40. Li, Q., Ma, Y., Oganov, A.R., Wang, H.B., Wang, H., Xu, Y., Cui, T., Mao, H.-K., and Zou, G. (2009) Superhard monoclinic polymorph of carbon. *Phys. Rev. Lett.*, **102**, 175506.
41. Mao, W.L., Mao, H.K., Eng, P.J., Trainor, T.P., Newville, M., Kao, C.C., Heinz, D.L., Shu, J., Meng, Y., and Hemley, R.J. (2003) Bonding changes in compressed superhard graphite. *Science*, **302**, 425–427.
42. Dellago, C., Bolhuis, P.G., Csajka, F.S., and Chandler, D. (1998) Transition path sampling and the calculation of rate constants. *J. Chem. Phys.*, **108**, 1964–1977.
43. Leoni, S. and Boulfelfel, S.E. (2010) Pathways of structural transformations in reconstructive phase transitions: insights from transition path sampling molecular dynamics. **This volume**.
44. Boulfelfel, S.E., Oganov, A.R., and Leoni, S.. (2010) *On the structure of "superhard graphite"*. In prep.
45. Oganov, A.R., Schön, J.C., Woodley, S.M., Tipton, W.W., and Hennig, R.G. (2010) Blind test of inorganic crystal structure prediction methods. **This volume**.

1
Periodic-Graph Approaches in Crystal Structure Prediction
Vladislav A. Blatov and Davide M. Proserpio

1.1
Introduction

The explosive growth in inorganic and organic materials chemistry has seen a great upsurge in the synthesis of crystalline materials with extended framework structures (zeolites, coordination polymers/coordination networks, metal–organic frameworks (MOFs), supramolecular architectures formed by hydrogen bonds and/or halogen bonds, etc.). There is a concomitant interest in simulating such materials and in designing new ones. In this respect, the role of new topological approaches in the modern crystallochemical analysis sharply increases compared to traditional geometrical methods that have been known for almost a century [1, 2]. As opposed to the geometrical model that represents the crystal structure as a set of points allocated in the space, the topological representation focuses on the main chemical property of crystalline substance – the system of chemical bonds. Since this system can be naturally described by an infinite periodic graph, the periodic-graph approaches compose the theoretical basis of the topological part of modern crystal chemistry. The history of these approaches is rather long, but only in the last two decades they have come into the limelight. Wells [3] was the first who thoroughly studied and classified different kinds of infinite periodic graph (net) and raised the question: what nets are important for crystal chemistry? This key question for successful prediction of possible topological motifs in crystals was being answered in two general ways initiated by Wells' pioneer investigations.

Firstly, mathematical basics for nets were developed in a number of works concerning quotient graph approach [4–10], special types of nets [11, 12], topological descriptors [13–16], and other topological properties related to nets, such as tiles and surfaces [17–19]. Secondly, net abundance and taxonomy were intensively explored in Refs. [20–26]. To solve the emerging problems, novel computer algorithms [5, 16, 19, 27, 28], program packages [6, 19, 27], and electronic databases [19, 27, 29] were developed that allowed to comprehensively analyze the topological motifs through hundred thousands of crystal structures. With these achievements, materials science and crystal chemistry come up to a new level of their development, that is, characterized by deeper integration of mathematical methods, computer

Modern Methods of Crystal Structure Prediction. Edited by Artem R. Oganov
Copyright © 2011 WILEY-VCH Verlag GmbH & Co. KGaA, Weinheim
ISBN: 978-3-527-40939-6

algorithms, and programs into modeling and interpretation of periodic systems of chemical bonds in crystals on the basis of periodic-graph representation.

The goal of this chapter is to show that the periodic-graph approaches are very fruitful not only to describe the crystal topologies, but become powerful tools to foresee possible topological motifs, to select most stable ones, and to design novel extended architectures. Besides the achievements of this new field of materials science that we may consider as the theoretical background of the so-called *reticular chemistry* [30], we will analyze the crucial problems that emerged after its rapid development in the last years.

1.2
Terminology

Since the subject of this chapter is not yet familiar to crystal chemists and materials scientists, we start with a brief summary of the specific terminology used. More detailed set of relevant definitions and the nomenclature of periodic graphs were given in Ref. [31].

Graph is a set of vertices (points), on which a *topology* is given as a set of ordered pairs of the vertices; each pair determines an edge of the graph. In chemistry, the graph vertices and edges correspond to atoms and interatomic bonds, respectively. The concept of topology is often used by chemists too broadly, without putting a strict sense into it [32]. We emphasize that when treating any molecular structure (crystal can be considered as an infinite molecule), "topology" or equivalently "topological structure" means nothing but the set of all interatomic bonds.

Net is a special kind of *infinite* graph, that is, *simple* (without loops, multiple, or directed edges) and *connected* (any pair of vertices in the graph is connected by a chain of edges); vertices of the graph are called *nodes* of the net. The *coordination number or degree* of a node is the number of edges incident on the node. A *subnet* (*supernet*) of the net A is a net whose sets of nodes and edges are subsets (supersets) of corresponding sets of A. Two nets are *isomorphic* if there is one-to-one mapping between the sets of their nodes and edges. The symmetry of the net is described by an *automorphism group* that enumerates all possible permutations of nodes resulting to isomorphic nets. Net is *n-periodic* if its automorphism group contains a subgroup being isomorphic to a group composed by n independent translations. *Embedding* of the net is a method of allocating its nodes in the space. A net embedding has *collisions* if it contains coinciding nodes and has *"crossings"* if some edges intersect. Two embeddings of a net that can be deformed into each other are *ambient isotopic*. Note that at no point in the deformation may edges intersect or have zero length. Clearly ambient isotopy implies isomorphism (but not *vice versa*). The symmetry of the net embedding can be lower (but not higher) than the symmetry of the net. For instance, the automorphism group of a net corresponding to a low-symmetric polymorph describes the high-symmetric phase as well. Moreover, there are *noncrystallographic* nets, whose maximal symmetry cannot be described by a crystallographic space group since any embedding of such

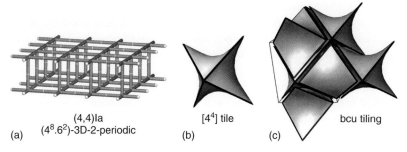

(a) (4,4)Ia $(4^8.6^2)$-3D-2-periodic (b) $[4^4]$ tile (c) bcu tiling

Figure 1.1 (a) Three-dimensional two-periodic net corresponding to the sphere packing (4,4)Ia (see [34] for terminology), (b) a single tile, and (c) the tiling for body-centered lattice (**bcu**). Hereafter we use *RCSR* bold three-letter symbols [29] to designate the net topology.

net has a lower symmetry. *Coordination figure* of the node is the solid formed by nodes incident to this node. *Dimensionality* of the net is equal to the dimensionality of the (Euclidean) space to which the net can be embedded without collisions. We emphasize that the net dimensionality can be larger than the net periodicity; for instance, 3D net can be two-periodic (Figure 1.1). The net is *uninodal (bi-, tri-, …, polynodal)* if all its nodes are equivalent (or there are two, three, …, many inequivalent nodes). If all edges of the net are equivalent, it is called *edge-transitive*. The net is *n-regular* if the degree of all nodes is equal n, even if the nodes are inequivalent. A *cycle (circuit)* is a closed path beginning and ending at a node, characterized by a size equal to the number of edges in the path (three-circuit, four-circuit, and so on). A *ring* is an n-membered cycle that represents the shortest possible path connecting all the $(n(n-1)/2)$ pairs of nodes belonging to that circuit. *Entangled* nets have independent topologies (they have no common edges), but they cannot be separated in any of their embeddings [33]. If the entangled nets have the same dimensionalities and the dimensionality of their array coincides with dimensionality of the separate net, they are called *interpenetrating nets*. Nodes of one interpenetrating net occupy the cages of the other net, and the edges of one net cross-cycles of the other net. *Tile* is a generalized polyhedron that can contain vertices of degree 2 and curved faces (that are rings of the net); it corresponds to a topological cage in the net. Tiles form normal (face-to-face) space partition, *tiling* (Figure 1.1). While any tiling carries a net formed by vertices and edges of tiles, the opposite is in general not true – not any net admits tiling. At the same time, most of nets admit an infinite number of tilings; there is a set of rules [28] to choose a unique, so-called natural tiling. *Dual* net has nodes, edges, rings, and cages corresponding to cages, rings, edges, and nodes of the initial net; the net and its dual net interpenetrate; if the net is equal to its dual net it is referred to as *self-dual*. If in this case the vertices of the dual net conform to the cages of the natural tiling, the net is called *naturally* self-dual. The net with p inequivalent nodes, q inequivalent edges, r inequivalent rings, and s inequivalent cages has the *transitivity pqrs*; often a shortened symbol pq is used especially if the tiling cannot be constructed. The transitivity of the tiling of a dual net corresponds to *srpq*. *Edge net* can be constructed by placing new nodes in the middle of the edges of the initial net, connecting new nodes by new edges and removing old nodes and edges. The edge net is *complete* if all the edges in the initial net are centered by new nodes and all old nodes and edges are removed; otherwise the edge net is *partial*. Similarly to edge net, one can construct a *ring net* by putting nodes in the centers of rings of the initial net, removing the nodes of the initial net belonging to the centered rings, and adding new edges between the centers of adjacent rings. Depending on whether all rings are centered or not, the ring net can be complete or partial.

For crystal structure prediction, the most significant nets are one-, two-, and three-periodic, which describe the topology of chain, layer, and framework crystal architectures, respectively. The nets of a higher periodicity as well as nets, whose automorphism group does not contain a subgroup isomorphic to a group of translations, have not yet been explored in relation to chemical objects, albeit they could be useful for aperiodic crystals.

Labeled quotient graph (*LQG*) of the net is a finite graph whose vertices and edges correspond to infinite sets of translation-equivalent nodes and edges of the net; it describes the net up to isomorphism. In general, LQG can have multiple edges and loops (Figure 1.2). Being finite, it is important for computer storage and processing of nets. The theory of LQGs is intensively being developed by Eon, Klee, and Thimm [7–9].

Topological index of the net is a set of numbers that characterize the net topology. The most rigorous topological index is *adjacency matrix*, that is, a quadratic matrix

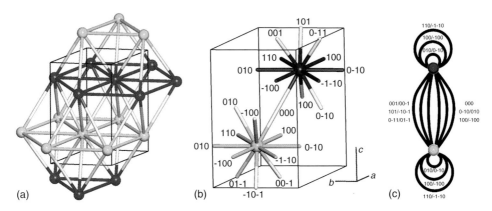

Figure 1.2 (a) A fragment of **hcp** net (hexagonal close packing) with the translation-equivalent nodes of the same shade of gray. (b) The two nodes in the primitive cell form the vertices of the LQG. (c) The links to the same translation-equivalent nodes (black-to-black, gray-to-gray) correspond to the loops in the LQG, while the links between the two nodes correspond to the sextuple edge of the quotient graph. The three-integer labels *uvw* show which translation-equivalent nodes are connected with the corresponding node of the net in the primitive cell. Depending on the direction chosen on the LQG, there are always two possible labels *uvw* or its centrosymmetric −*u*-*v*-*w*: for example, going out of the cell along 100 is equivalent to enter from −100.

M, whose rows and columns correspond to vertices of the LQG, and each entry M_{ij} shows the existence of edge between ith and jth vertices. The adjacency matrix defines the net and its LQG up to isomorphism. Some other kinds of topological indices are useful to identify and compare net topologies [13–16, 35].

1.3
The Types of Periodic Nets Important for Crystal Structure Prediction

Since there is an infinite number of topologically different periodic nets and an infinite number of geometrically different embeddings of a particular net, it is important to determine the types of nets to be most important for predicting topological features of crystal structures. This work has mainly been done by O'Keeffe and coworkers during the last two decades [21, 36, 37]; it still proceeds now. Three-periodic nets were studied in much more details than low-periodic ones (see Section 1.6); the special types of nets are described below with the example of three-periodic nets.

The easiest to enumerate are the most symmetrical topologies that have the smallest number of inequivalent nodes and/or edges of the net. Hence, *uninodal* and *binodal* nets as well as *edge-transitive* nets are of special interest for crystal chemistry. Among them, one can separate five *regular* nets: **srs**, **nbo**, **dia**, **pcu**, and **bcu** (Figure 1.3), where the coordination figures are regular solids (triangle, square, tetrahedron, octahedron, and cube, respectively), one *quasiregular* net (**fcu**) with two kinds of faces in the coordination figure (cuboctahedron), and 14 *semiregular* nets that are both uninodal and edge-transitive. Of other nets, *n-regular* nets and *minimal* nets (for the latter holds that removing any set of equivalent edges gives rise to decrease of the net periodicity) are important to crystal structure prediction [12, 37, 38].

In chemical compounds, the atoms tend to keep neighboring atoms at similar distances; as a rule, the difference in distances for a particular pair of atoms does not exceed several tenths of angstrom. In this respect, *sphere packings* are most important: they describe net embeddings with all equal edges corresponding to the shortest distances between nodes, only the first neighboring nodes are connected, all other nodes are placed at larger distances. At present, all uninodal sphere packings are known for one- and two-periodic nets [34]; the search for three-periodic uninodal nets is almost finished [39, 40, and references therein]; at least we know all three-periodic uninodal sphere packings that emerge in crystals.

We also mention a special group of *self-dual* nets, that is, closely related to interpenetrating structures; since the nodes of one net occupy the cages of another net in the interpenetrating array, the net isomorphic to its dual net will be the most suitable to form such an array. At present, 13 naturally self-dual nets are listed in the reticular chemistry structure resource (*RCSR*) database [29]: **cds**, **dia**, **ete**, **ftw**, **hms**, **mco**, **pcu**, **pyr**, **qtz-x**, **sda**, **srs**, **tfa**, and **unj** (Figure 1.4). Possible types of interpenetration of homogeneous sphere packings were derived in Ref. [11]; the

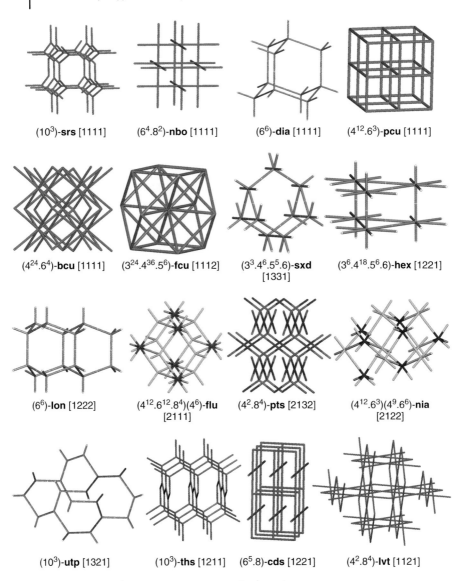

Figure 1.3 List of some important nets. Each net is characterized by its point symbol, RCSR three-letter name and transitivity [29].

classes of interpenetration were considered [20, 22] in relation to types of symmetry operations relating the nets in the array.

An efficient way to find crystallochemically important nets and to generate new ones is to search for subnets of known nets. Blatov and Proserpio [16, 41] obtained all uninodal and binodal subnets (totally more than 48 000) for almost all known nets of the types described above. But not less important is to generate nets *ab initio*,

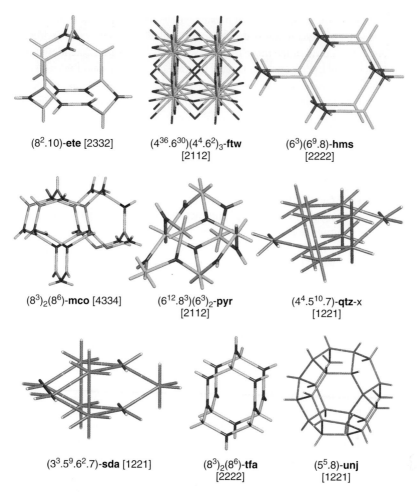

Figure 1.4 Nine self-dual nets. For **srs**, **cds**, **dia**, and **pcu** see Figure 1.3.

independently of peculiarities of crystal structures. Analyzing such nets, one can see which of their features prevent them from realizing in nature. A very fruitful method was proposed by Hyde and coworkers [19, 42] to obtain three-periodic nets (*epinets*) by projecting 2D hyperbolic tilings.

1.4
The Concept of Topological Crystal Structure Representation

The question "what net corresponds to a given crystal structure" is not so easy as it likely seems. Any crystal structure can be represented in different ways depending on which interatomic interactions and structure groups we take into account. For instance, we can consider the crystal structure of an ice polymorph as a packing of water molecules, as a net of water molecules connected by hydrogen bonds or as a net of oxygen and hydrogen atoms linked by valence and hydrogen bonds. Of course, some other more or less exotic variants can be considered, for example, packing of hydrogens, but usually they are chemically meaningless. How to enumerate all possible topologies and select reasonable ones? Thus, we come to the concept of *topological crystal structure representation* [43], that is, defined as a net generated from the *complete* net of the crystal structure by some method of its simplification. By complete net, we mean the net that includes all atoms of the structure as the nodes and all possible, even weakest, atomic interactions as the edges; the *complete* crystal structure representation corresponds to this net. This hypothetical net describes the structure topology in all details, but in any particular crystallochemical investigation we have to simplify it somehow to obtain a *partial* representation. The following simplification procedures with the complete net enable us to produce any structure representation: (i) removing some edges; (ii) removing some nodes together with the incident edges; (iii) contracting some nodes to other nodes or to some other points (commonly the molecular centers of mass, the barycenter) keeping the net connectivity.

The first simplification procedure is explicitly or implicitly used in any description of crystal structure. Ordinarily, we consider only strong interactions like valence, ionic, or metallic bonding and ignore van der Waals or even weaker interactions (e.g., Coulomb interactions between long-distant ions). We apply the second simplification procedure when, for instance, interstitial ions or molecules are omitted in zeolites or MOFs. The last simplification procedure corresponds to the representation of polyatomic groups as structureless particles, when we analyze packings of inorganic ions in complex salts or organic molecules in molecular crystals. Most crystal structure representations are the result of a combination of the simplification procedures. For example, one can treat the ice VIII polymorph as an assembly of hydrogen and oxygen atoms connected by valence and hydrogen bonds applying only the first simplification procedure to omit other weak H···O and H···H interactions, which results in the SiO_2-cristobalite topology of the corresponding net. Moreover, there are two such equivalent interpenetrating nets in the structure like in Cu_2O. The next way is to consider the structure as a net of water molecules connected by hydrogen bonds. To provide this representation, one has to contract/remove hydrogen atoms to oxygens, which gives rise to the diamond (**dia**) topology and, hence, to the array of two interpenetrating **dia** nets (Figure 1.5). At last, the contraction procedure can be applied with accounting all interactions H···O and H···H between neighboring molecules to obtain the single net of the body-centered cubic (**bcu-x**) topology corresponding to molecular packing, where each water molecule contacts 14 other molecules (Figure 1.5). More examples of different crystal structure representations are given in Section 1.7.2.

As a result of simplification, we obtain an *underlying net* that elucidates the general topological motif of the crystal structure. It follows from the aforesaid that several underlying nets can correspond to the same structure depending on the simplification method.

1.4 The Concept of Topological Crystal Structure Representation

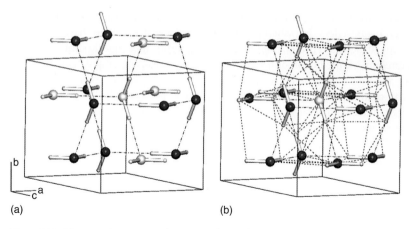

(a) (b)

Figure 1.5 The crystal structure of ice VIII polymorph: (a) as an array of two interpenetrating diamondoid **dia** nets; (b) as a single **bcu-x** net; the environment of a water molecule with 14 other molecules is shown. Hydrogen and van der Waals bonds are shown as dash-dotted and dashed lines, respectively.

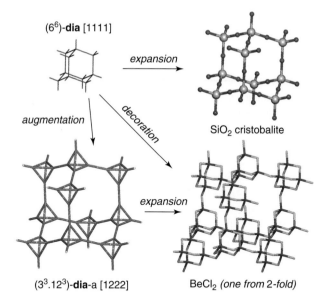

Figure 1.6 Examples of decoration, augmentation, and expansion transformations.

The transformations, being inverse to the simplification procedures, are the *decoration* of the net nodes (with the special case of *augmentation* in which the nodes of the original net are replaced by a group of vertices with the shape of the original coordination figure of the node) [29] and the *expansion* of the net edges; they are the main operations of reticular chemistry [30, 44] (Figure 1.6).

Thus, the question, what is the single, "true" topology for a particular compound, is often meaningless, but when predicting the crystal structure we have to foresee as many structure representations as possible. From topological point of view, a comprehensive prediction of the crystal structure means finding its complete net since all other representations can be derived from the complete one. In practice, we can predict only some partial representations. For example, we can state that most of molecular packings tend to have topologies of the close packings fcc or hcp; however, we cannot often predict how the atoms of molecules are allocated in the space. Even having a comprehensive X-ray information on the molecule geometry and environment, sometimes we cannot prove if some interactions exist or not. The concept of crystal structure representation allows us to realize which level of the structure understanding we have accomplished and what other representations can be derived from this one.

1.5
Computer Tools and Databases

New tasks emerged in crystal chemistry at the end of the 20th century invoked new tools for their solution. Since the solution should rely on a great deal of crystallographic data, the tools had to provide electronic processing of the information. At that time, just about 10 years ago, there were worldwide electronic databases that accumulated general crystallographic information on almost all studied crystal structures, but these databases were used mainly as electronic handbooks, not as tools to search for crystallochemical regularities or to predict novel crystal architectures. Principally, new software should be developed to automate main stages of crystallochemical analysis and to process the databases in a batch mode. This software naturally uses the periodic net concept to represent and explore the crystal structure. Now we have the following program packages that are distributed free of charge:

Gavrog[1] includes two programs: *Systre (Symmetry, Structure Recognition, and Refinement)* [6] finds the net embedding with maximal symmetry and provides the topological classification using a built-in archive of more than thousand nets taken mostly from the *RCSR* database (see below); *3dt (3D tiler)* visualizes three-periodic tilings [37] and computes their topological parameters.

Olex[2] [45] provides, besides some standard procedures of structure determination and analysis, the tools for topological simplification of the initial periodic net and computing a number of its topological indices.

TOPOS[3] [27] accumulates all topological approaches mentioned in the previous parts and provides the study of both individual crystal structures and of large groups of structures stored in electronic databases. Almost all comprehensive investigations of net occurrence discussed in Section 1.6 were performed with

1) *Generation, Analysis, and Visualization of Reticular Ornaments using Gavrog*; http://www.gavrog.org.
2) http://www.olex2.org.
3) http://www.topos.ssu.samara.ru.

TOPOS. *TOPOS* is integrated with databases on topological types *TTD* and *TTO* described below.

Using the topological software we can determine a number of crystal structure parameters and perform some operations to be important for the structure classification, rationalization, and prediction: (i) topological type of the underlying net (including the type of interpenetration, if any) and local topology of complex group; (ii) maximum-symmetry embedding for new nets; (iii) automatic search for the net representations; (iv) automatic simplification of the net; (v) search for supernet–subnet relations; (vi) search for a given finite fragment in the net; (vii) performing all the operations both for single crystal structure and for large number of structures *via* interface with crystallographic databases. The program packages contain tools to discover general tendencies in formation of topological motifs; many hypotheses, or models based on the periodic net concept can now be rapidly and comprehensively checked. Examples will be considered in Sections 1.6 and 1.7.

Gavrog Systre and *TOPOS* were used to create a new type of electronic databases that can be called *crystallochemical* since their main information is related to the topology of periodic nets of chemical bonds as opposed to *crystallographic* databases that are focused on the arrangement of atoms in the space. To create a record in a crystallochemical database, one has to restore a periodic net from the crystallographic data, simplify it if required, reduce the resulting infinite net into an LQG, and store the adjacency matrix of this graph as well as other its topological indices. This can be done only with the tailored software described above. Currently, the following free crystallochemical databases are developed:

RCSR[4] [29] is the oldest one, but of age less than 10 years. It contains maximum symmetry embeddings of more than 1700 two- and three-periodic nets considered important for crystal chemistry and crystal design. Most of the nets were found in crystal structures, the remaining ones can be considered as suitable templates for new materials. One of the reasons to decide in favor of a particular crystal structure representation is if its topology is found among the *RCSR* nets. *RCSR* uses three-letter symbols to denote the net topologies; these symbols are used in this chapter.

EPINET[5] project announced in 2005 [19, 42], strictly speaking, cannot be considered as a crystallochemical database since it collects the nets generated *ab initio*, irrespective of crystal structures. However, it is strongly important to interpret the results of net occurrence and to develop the methods of net topology prediction as will be shown in Section 1.6. Now *EPINET* includes 14 532 epinets, 162 of them coincide with *RCSR* nets.

TTD and *TTO* collections[6] [41] form an integrated set; the *TTD* part contains the information on topological indices of almost 68 000 topological types of nets including *RCSR* and *EPINET*, and the *TTO* part collects the links between the topological types and the crystal structure data stored in crystallographic databases.

---

4) *Reticular Chemistry Structure Resource*, http://rcsr.anu.edu.au/.
5) *Euclidean Patterns in Non-Euclidean Tilings*, http://epinet.anu.edu.au/.
6) *TOPOS Topological Databases* and *Topological Types Observed*, http://www.topos.ssu.samara.ru.

TOPOS uses this set to determine crystal structure topology and to find other topologically similar compounds.

What information useful for the prediction of topological motifs can we extract from the topological databases? Resting upon the topological type of the net, we can immediately draw some conclusions. (i) If the type is stored in *RCSR*, this topology is not occasional; it was already considered due to some of its important peculiarities. If the net is absent in the databases, there are some special reasons that have invoked the net and we have to discover them. (ii) Maximum symmetry embedding including symmetry of nodes and coordination figures is useful to apply the symmetry criteria discussed in Section 1.7.1. (iii) If the net has some special features (see Section 1.7), the corresponding topological motif could be particularly stable. (iv) Knowing other structures with the same topology, we can find similarities between the compounds and materials of different composition, chemical nature, and structure details that are inessential in the given topological representation. But the most important point is that the tools described in this part allow us to solve currently the crucial problem of crystal chemistry, the nonuniform abundance of periodic nets in nature.

1.6
Current Results on Nets Abundance

The question why some topological motifs frequently occur in nature while other nets have been never found is a crucial point in understanding how the resulting crystal architecture depends on chemical composition and bonding in the substance. Quantum-mechanical technique can hardly be applied as a universal tool to properly answer this question since different topological models of the same structure often lie too close to each other on the energy landscape [46, 47]. Hence, some empirical regularities based on geometrical and topological properties of the structures should be discovered. Being less general than the quantum-mechanical approach, they have the advantage to be drawn and verified with the whole immense array of experimental crystallographic data. The first step in solving this problem is to build statistical distribution of the net topologies over all known periodic structures. This task is very time-consuming, but recent progress in computer methods of the net analysis allows us to hope that it will be finished quite soon. Moreover, even current results on nets occurrence can provide some conclusions to be important for successful topology predictions.

The first extensive investigation of net occurrence was reported in Ref. [20] for interpenetrating valence-bonded 3D architectures. Using *TOPOS*, the authors processed the whole CSD and obtained a comprehensive list of 301 interpenetrating coordination polymers. It was found that the three most preferred topological motifs are **dia**, **pcu**, and **srs**. Just after a year, O'Keeffe and coworkers [21] published the statistics for 774 valence-bonded single nets in MOFs with the same sequence of the first three leaders. In the next 4 years, the crystal structure topologies in organic, inorganic, and metal–organic compounds were systematically investigated

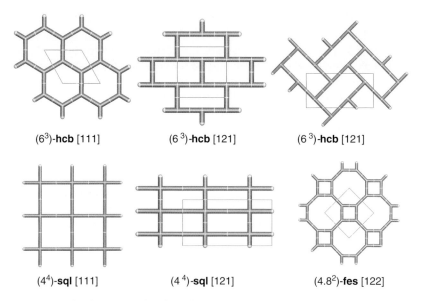

Figure 1.7 The three most abundant plane nets **hcb**, **sql**, **fes** with some low symmetry embedding (italicized).

[22–26, 48]; most results of these studies are collected in the *TTO* databases. Thus, in valence-bonded interpenetrated inorganic and MOFs as well as in single MOFs, the leaders are **dia–pcu–srs**; in organic molecular crystals with hydrogen-bonded single networks, the results are similar (**dia–pcu–sxd–hex**) as well as for hydrogen-bonded coordination compounds (**pcu–bcu–hex–dia**).

It seems strange, but all published statistics concern only three-periodic structures of organic, inorganic, and coordination compounds; the low-periodic nets were not explored. At the same time, in two-periodic motifs the preference of some topological types is much sharper than in three-periodic structures. Thus, our analysis of 1711 two-periodic coordination networks shows that the first three places are occupied by **sql** (37.2%), **hcb** (17.3%), and **fes** (7.0%) plane nets (Figure 1.7). Moreover, if we consider only the contribution for the nets of the same coordination, it becomes much more expressed: **hcb** and **fes** compose 65.1% and 26.2% of three-coordinated nets, respectively, and **sql** covers 88.6% of four-coordinated nets. Similar results are obtained for interpenetrating two-periodic arrays. Thus, for 271 coordination networks containing valence-bonded interpenetrating layers, the most abundant layer topologies (84%) are **sql** (59.4%), **hcb** (24.0%), and **fes** (4.8%).

As was mentioned above, besides the analysis of crystal structures of chemical compounds, not less important is the consideration of artificial nets generated by some tailored methods. This approach allows one to extend the list with the nets that never occur in nature. The database *EPINET* mentioned above is the greatest project in generating *ab initio* nets. Studying occurrence of epinets, one can realize how stochastic is the realization of topological motifs in crystals. Therefore, since publishing the first *EPINET* release in 2005, the epinets were considered in any review mentioned above. Summarizing the results, we see that only 192 out of 14 532 epinets (1.3%) have examples in nature; this fact proves the occurrence of nets to be strongly nonrandom. One more method for generating nets was developed by Blatov [16]; it is based on deriving all subnets for a particular net. Using this method for the *RCSR* nets, Blatov and Proserpio [16, 41] generated 9988 uninodal and 38 304 binodal nets that were never described before; only for 92 and 100 of them, the crystal structure examples have so far been known. Resuming, the probability of a randomly taken topology to be realized in crystals is not larger than 1–2%.

The current results on nets occurrence reveal general tendencies that require a theoretical interpretation: (i) there are a few nets that are much more frequent than others; (ii) diversity of single nets is much larger than of interpenetrating arrays; (iii) although bond type influences the occurrence of nets, some nets, like two-periodic **sql**, **hcb**, and **fes** or three-periodic **srs**, **dia**, and **pcu**, are most important for the substances of different nature; and (iv) the variety of nets directly depends on diversity of coordination centers; topology of nets with one kind of node only is the easiest to predict.

This analysis raises the questions to be crucial in the crystal structure prediction: (i) why frequent topologies are frequent and rare topologies are rare? (ii) Can any net type form an interpenetrating array, or some nets can occur single only? (iii) Can we predict topology for a given substance, with what probability? And what parameters should we know? Although these questions are not yet completely answered, there is an obvious progress in understanding as shown in detail in the next part.

1.7
Some Properties of Nets Influencing the Crystal Structure

1.7.1
Symmetry of Nets and Embeddings

The symmetry of a crystallographic periodic net can be described in terms of space groups; therefore, one can expect that the well-elaborated technique of mathematical crystallography can be used to derive some general properties of nets. Here we collect the results in this field with respect to prediction of the target motifs.

Theorem 1.1. *The space group of a net is a subgroup of its automorphism group.* This means that the symmetry owned by the net can be higher than the space-group symmetry of any of its embedding. If it is so, the net is noncrystallographic [7] and can have embeddings with space groups being not in a group–subgroup relation, for example, both in cubic and in hexagonal symmetry. Noncrystallographic nets always have symmetry operations that are inconsistent with a given space group. One can expect that these operations can appear as a kind of supersymmetry, and the substances built with such nets can have some special properties. However, nobody has yet studied the occurrence of noncrystallographic nets since there are no clear criteria to detect them in crystal structures.

Corollary 1.1. *Site symmetry group of the net embedding is a subgroup of the symmetry group of a node in the net.* This corollary restricts possible local topologies of structural units (i.e., both complex groups in the initial structure and the resulting coordination figures in the simplified net); for instance, the units with hexagonal symmetry are not suitable to realize in a cubic symmetry since they are forced to be distorted.

Corollary 1.2. *The point group of a structural unit in the crystal must be a common subgroup of the point group of the corresponding node in the most symmetric embedding of the net and the point group of the isolated structural unit.* A good illustration of this corollary is the consideration of possible topologies for molecular crystals. In particular, it follows that a uninodal net of a given topology may be realized in some space-group symmetry G' if, and only if, the index of G' in G is a divisor of the order of the point group of the node in the most symmetric embedding of the net [49, 50]. This conclusion is proved by a comprehensive analysis of molecular packings [23, 26].

The theorem provides many useful conclusions to predict topological properties of the crystal architectures. In general, the specified topology restricts possible geometrical embeddings of the net, and *vice versa*. We can now answer the following crystallochemical questions [23, 26]:

1) *Can a given topology be realized in a given space group?* For instance, the diamond (**dia**) net cannot have hexagonal symmetry. Indeed, the automorphism group of **dia** is isomorphic to the space group Fdm, but any hexagonal space group is not a subgroup of $Fd\bar{3}m$ that contradicts Corollary 1.1. The same concerns the lonsdaleite (**lon**) net (so-called hexagonal diamond) that cannot have $Fd\bar{3}m$ or any other cubic symmetry since its most symmetric embedding has the space group $P6_3/mmc$.
2) *Which space groups can be realized for low-symmetric embeddings of a given topology?* For this purpose, we have to find the most symmetric embedding of the net with the space group G; the space group of any other embedding of the net will be a subgroup of G. For example, both **dia** and **lon** can have embeddings with orthorhombic $Pnna$ symmetry, that is, a common subgroup of $Fd\bar{3}m$ and $P6_3/mmc$.
3) *Can the coordination figure of a given symmetry be realized in a given embedding?* For example, can we obtain **dia** with square-planar coordination? In the most symmetric embedding, any node of the diamond net has the symmetry $\bar{4}3m$ (T_d). According to Corollary 1.1, we cannot obtain the regular square-planar coordination since the point group of the square ($4/mmm$ or D_{4h}) is not a subgroup of $\bar{4}3m$. The same concerns rectangular coordination with the point group mmm (D_{2h}). However, we can find some distorted planar coordination figures without inversion, describing by a subgroup of $\bar{4}3m$, say, $mm2$ (C_{2v}).

4) *Can a given structure unit occupy a particular Wyckoff position in a given net embedding?* If we know the symmetry of the net embedding as well as the symmetry of a molecular group, we can use Corollary 1.2 to decide if the symmetries are compatible with each other. For example, molecules with inversion center prefer not to occupy the positions in the diamondoid nets because they have to lose the inversion center. The analysis of molecular packings [23, 26] proves this rule.

5) *Will a given molecular packing in a given space symmetry be monomolecular or not?* For instance, a uninodal diamond net is forbidden in the $P2_1$ group because its index in $Fd\bar{3}m$, $i = 48$, is not a divisor of the order of the point group $\bar{4}3m$ of the node ($i = 24$), so at least two inequivalent nodes (molecular centers) should exist in the unit cell of a $P2_1$ diamond [23]; the total order of these nodes ($24 \times 2 = 48$) is equal to i and, hence, i is its divisor that obeys Corollary 1.2.

In all the applications of the symmetry relations, the crucial is the space-group symmetry of the net in the most symmetric embedding. Once this symmetry has been found, the technique of mathematical crystallography can be in full applied to the net. Unfortunately, no tools were proposed to determine the automorphism group of the net except the geometrical approach based on the barycentric placement (*Systre*). Therefore, the nets that have a barycentric placement with collisions (including noncrystallographic nets) remain difficult to be considered within this approach. But in our experience, such nets are extremely rare, so *Systre* provide the most symmetric embedding for almost all chemically relevant examples.

Less strict, but not less important assumption concerning significance of high-symmetric nets for crystal chemistry was proposed in Ref. [21]. The authors revealed that most frequent underlying nets in MOFs are the nets with *high space-group symmetry* as well as *high site symmetry* of the nodes. It is noteworthy that the symmetry of the crystal can be and usually is low, but the symmetry of the underlying net itself ordinarily is high. A useful criterion of the high symmetry is transitivity. According to Ref. [21], the most important nets are regular, with transitivity 1111, that are **srs**, **dia**, **nbo**, **pcu**, and **bcu**. However, all nets with one kind of node and edge, that is, with transitivity 11*rs*, are also suitable to be observed. This assumption explains why the most frequent nets in three-, four-, and six-coordinated MOFs are **srs**, **dia**, and **pcu**, respectively. The authors [21] substantiate the high-symmetry criteria by isotropy of the reacting system (melt or solution) and reaction centers (metal atoms). The reason could also be that the underlying net corresponds to some "primary" structure motif that determines the general method of ordering (the main modes in phonon spectra), while the details of interactions between structural units provide geometrical distortions of the structure and result in some subgroup of the space group of the underlying net. In any case, the physical reasons of this phenomenon require a more thorough consideration.

The role of high symmetry of the most symmetrical embedding of the net invokes some special classes of nets to be important for structure prediction [37].

1) *Regular* (transitivity 1111), *quasiregular* (transitivity 1112), and *semiregular* nets (transitivity 11rs) all have one kind of vertex and edge but differ by coordination figures that correspond to regular, quasiregular, or arbitrary convex polyhedra, respectively. All of them occur in crystal structures not occasionally [21, 37].
2) *Edge-transitive* nets have one kind of edge; in addition to the nets of class (1) the edge-transitive nets of transitivity 21rs can be important for structures with two or more kinds of structural groups. Such edge-transitive nets as **flu**, **pts**, **nia** are frequent in MOFs [21, 51, 52].

1.7.2
Relations Between Nets

Starting from a particular net, we can produce other nets that are unambiguously related to the initial one. Let us consider the types of such nets to be most important for materials science.

1) *Subnets.* One more important application of symmetry properties of periodic nets is the enumeration of all shortest ways to transform one net to another. For example, one can be curious of how to transform diamond to lonsdaleite by breaking and forming the minimum number of bonds. The space group of the intermediate net, supernet, or subnet must be a common subgroup of the space groups $Fd\bar{3}m$ and $P6_3/mmc$; moreover this net must be a common supernet or subnet of **dia** and **lon**. A corresponding algorithm was proposed and realized in Ref. [16]. Using the program package *TOPOS*, one can construct a so-called net relation graph that indicates supernet–subnet relations for a given set of nets. With the net relation graph, we find that the pair **dia–lon** has two common three-coordinated uninodal subnets: **utp** (*Pnna*) and **ths** ($I4_1/amd$). Since *Pnna* is a common subgroup of $Fd\bar{3}m$ and $P6_3/mmc$, **utp** can exist in its highest symmetry during the transition, while the possible symmetry **ths** is $C2/c$, that is, the maximal common subgroup of $Fd\bar{3}m$, $P6_3/mmc$, and $I4_1/amd$ that retains **ths** uninodal. The number of pathways through a five-coordinated common supernet is much higher: according to Ref. [16], there are 93 such pathways. This information could be useful to predict structure transformations during phase transitions.
2) *Dual nets.* Since the nodes and edges of the dual net correspond to cages and rings of the initial net, the dual net can be considered as a system of void centers and channel lines between them. Anurova and Blatov [53] showed that dual nets can be efficiently used to analyze the migration paths in fast-ion conductors, to predict the existence and dimensionality of conductivity. In MOFs, the self-dual nets are of special interest [12] because, being combined with the initial net, they form strongly catenated network arrays. Thus, the uninodal self-dual nets **srs**, **dia**, **cds**, and **pcu** compose the main part (70%) of interpenetrating MOF arrays [20]. Self-dual two-periodic nets can easily form a stacking, where nodes of one net are projected to the centers of rings of

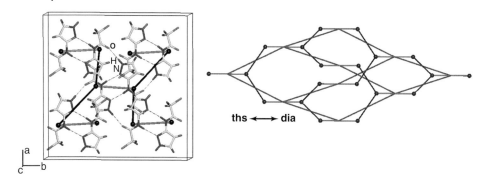

Figure 1.8 (a) The crystal structure of 3-(chloroacetamido)pyrazole with the **ths** underlying net and (b) corresponding partial edge net of the **dia** topology with the four-coordinates nodes on the middle of the gray edges. Molecular centroids and H-bonded dimers (R_2^2 (8) synthon [24, 25]) are shown in the initial net as black balls and gray lines.

another net [54]. This feature provides existence of polytype series like the infinite series of close packings.

3) *Edge nets.* This type of nets can be useful for the design of molecular crystals. For instance, the net, whose nodes correspond to dimers, is an edge net with respect to the net of molecular centroids. Thus, both **ths** and **srs** nets have partial edge nets of the **dia** topology (Figure 1.8). Some other relations are given in Ref. [23].

4) *Ring nets.* A natural application of ring nets is the description of topology of coordination polymers with synthons. The ring net describing synthons is partial, as a rule [23]. For instance, an **lvt** net has partial ring net of the **dia** topology (Figure 1.9). This means that the synthons in the **lvt** net can arrange in a diamondoid motif. In turn, the **lvt** topology describes a partial ring net for the **gis** atomic net, and so on.

1.7.3
Role of Geometrical and Coordination Parameters

The criteria of high symmetry are not sufficient to explain an important role of some "default" nets. For example, the net **sra** has rather low symmetry (*Imma*), not minimal transitivity (33), does not belong to the special classes mentioned above, and its nodes are not allocated in the most symmetrical positions in the space group (8*i* with site symmetry *m*). However, **sra** occupies the second place after **dia** in the list of the most frequent MOF topologies with tetrahedral units [21], while two nets following after it – **sod** (*Im$\bar{3}$m*) and **qtz** (*P6$_2$22*) – have much higher symmetries and the smallest transitivities (12). Obviously, additional geometrical and chemical factors should manage the net occurrence.

The geometrical parameters are the property of the net embedding; therefore, special types of embeddings, like sphere packings, should be specified and explored.

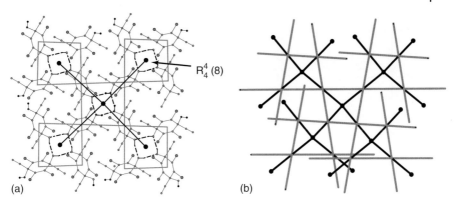

Figure 1.9 (a) The crystal structures of 2,3,5,6-tetrachloro-trans-1,4-diethynylcyclohexa-2,5-diene-1,4-diol (XEHKIE) with $R_4^4(8)$ synthons; the initial net (gray lines) is formed by the centroids of molecules, while the partial ring net (black balls and lines) describes the synthons; (b) initial net **lvt** (gray) and partial ring net **dia** (black).

The importance of sphere packings reflects the importance of edge-transitive nets; however, in this case the equivalence of edges is not necessarily caused by symmetry reasons. It is a typical requirement that the links in the structure should be not the same, but of similar lengths (e.g., inorganic frameworks of oxofluorides or coordination polymers with chemically equivalent, but conformationally mobile ligands). It is noteworthy that all the nets mentioned above, including **sra**, are sphere packings. So if we have a compound with the ligands similar by length, the resulting net will be a sphere packing. Otherwise we can obtain a sphere packing or not; it depends on other reasons.

At present, not all sphere packings are derived even for uninodal nets. The occurrence of the known sphere packings is quite different; there are some of them that have no examples in crystal structures. In some cases, it can be explained by their low symmetry, but, in general, a deeper exploration of their geometrical properties is required. A more detailed classification of sphere packings accounting their ability to distortion due to degrees of freedom was proposed in Ref. [12]. However, no comprehensive study has yet been performed to check the occurrence of nets with respect to this classification.

The next important parameter is the form of coordination figure of the node. Not only distances (that are crucial for the concept of sphere packing) but also angles should be taken into account when anticipating the underlying net topology; moreover, in some cases there is a strong dependence between coordination figure and the underlying topology. The best correlations are observed for two-periodic nets. For instance, in the analysis of coordination polymers square-planar coordination leads to the **sql** topology almost in all cases, even if the rectangular is not regular. The same concerns triangular coordination figure that corresponds to the **hcb** underlying topology. We emphasize that **hcb** and **sql** are not unique topologies with planar three- or four-coordinated nodes, but they are the only uninodal

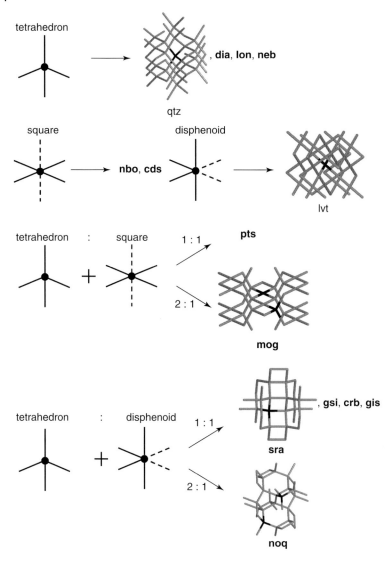

Figure 1.10 Tetrahedral, octahedral, and related complexes, coordination figures, and corresponding underlying nets in cyanides. Dashed lines correspond to links with terminal ligands. For some nets see Figure 1.3, for the others refer to the *RCSR* web page [29].

nets with triangular and square/rectangular coordination figures. Obviously, the highest possible plane-group symmetry plays a determinative role in this case. The required topology and geometry of the coordination figure can be designed by a proper introduction of terminal ligands into the coordination sphere of metal atom (Figure 1.10).

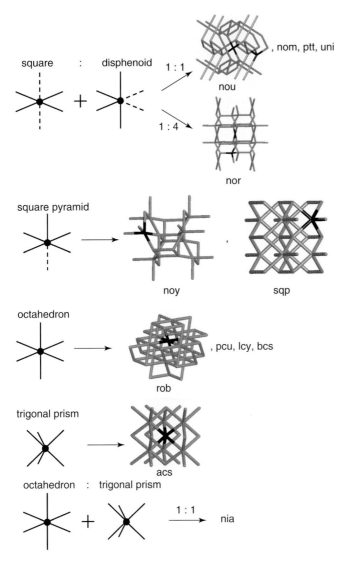

Figure 1.10 (continued.)

In the three-periodic coordination polymers, the correlation is not so strong; the same coordination figure can result in a diversity of underlying topologies [21]. However, if all ligands in the structure are similar by size, some special topologies of underlying net can be anticipated for any type of local coordination of metal atoms. For example, in cyanides there are the following correlations ([55], Figure 1.10):

- *Four-coordinated nodes with tetrahedral coordination figure* in almost all cases of cyanides result in **dia** topology; the only exception is **lon** topology, that is, very close to **dia** but is not edge transitive. One more edge-transitive topology with tetrahedral coordination, **qtz**, is realized only in inorganic cyanides. In this case, the spacers between tetrahedral centers are not cyanide ions, but linear groups $[Au(CN)_2]^-$ that, being more flexible provide less uniform quartz topology. More voluminous spacers, in particular, complex groups Cu(3,10-dipropyl-1,3,5,8,10,12-hexa-azacyclotetradecane) [56], lead to the **neb** topology, that is, locally very similar to **dia** [24].

- *Four-coordinated nodes with square-planar coordination figure* give two topologies, **nbo** and **cds**; the former one looks more preferable since it is edge-transitive (1111 vs. 1221).
- *Four-coordinated nodes with disphenoidal coordination figure*, which can be obtained from trigonal–bipyramidal coordination by removing one equatorial ligand or from octahedral coordination by removing two neighboring ligands, are realized in the mercury cyanide and correspond to the **lvt** topology. Obviously, the geometrical features of disphenoid (angles of 90° and 180°) provide formation of four- and eight-rings that are typical for **lvt**, despite different coordination figure in **lvt** (square). We are not aware of the uninodal nets with disphenoidal coordination in the maximal symmetry embedding.
- *Four-coordinated nodes with tetrahedral and square-planar coordination figures* give rise to two topologies – **pts** or **mog** – depending on the nodes ratio 1:1 or 2:1, respectively.
- *Four-coordinated nodes with tetrahedral and disphenoidal coordination figures* lead to γ-Si (**gsi**), CrB_4 (**crb**), $SrAl_2$ (**sra**), or zeolite gismondine (**gis**) topologies at the nodes ratio 1:1. In this series, the number of four-rings meeting at the node increases from 0 to 3. Probably, different distortion of the disphenoidal coordination figures could manage the resulting underlying topology. At the nodes ratio 2:1, only the **noq** topology is realized.
- *Four-coordinated nodes with square-planar and disphenoidal coordination figures* emerge when there are two octahedral centers where two terminal ligands are allocated in *trans*- or *cis*-positions, respectively. At the ratio 1:1 the topologies **nom**, **nou**, **ptt**, or **uni** are observed, while the ratio 1:4 gives rise to the **nor** underlying net.
- *Five-coordinated nodes with square-pyramidal coordination figure* are derived from octahedral complexes with the single terminal ligand. This coordination produces the **noy** topology in all cases. If voluminous bridges like Me_3Sn in $[(n\text{-}Bu_4N)_{0.5}(Me_3Sn)_{3.5}Fe(CN)_6]$ [57] link the coordination centers, the **sqp** topology is realized.
- *Six-coordinated nodes with octahedral coordination figure* and short (cyanide) bridges always correspond to the **pcu** underlying net. The exceptions arise when the bridges are long or/and voluminous. Thus, in $N(n\text{-}Bu)_4\{Ni[Au(CN)_2]_3\}$ [58], very rare **bcs** topology is realized. Although the net **bcs** is semiregular, it has strongly distorted coordination figure in the maximal symmetry embedding, since at the regular octahedral coordination two additional disconnected nodes are located nearby the central one (Figure 1.11a). In $N(n\text{-}Bu)_4\{Ni[Au(CN)_2]_3\}$, this becomes possible owing to the long $[Au(CN)_2]$ bridges; out of eight Ni atoms surrounding

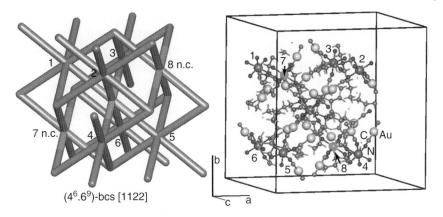

Figure 1.11 (a) Idealized **bcs** net with the six coordinates nodes (1–6) and the two close but not directly connected to the central one (7,8 n.c.); (b) a fragment of the N(n-Bu)$_4${Ni[Au(CN)$_2$]$_3$} structure. Ni atoms (1–8) form a cube around the central Ni. The two atoms disconnected to the central one (7,8) are marked by arrows.

the central Ni at distances 10.1–10.2 Å, only six are connected to it via [Au(CN)$_2$] groups (Figure 1.11b). In isomorphic pair {(n-Bu)$_3$ Sn}$_3$[M(CN)$_6$](M = Fe, Co) [59], the three-rings required for emerging the **lcy** topology are formed of voluminous [(n-Bu)$_3$ Sn] bridges. If the coordination center is extended with a binuclear cluster like [Cu$_2$(CN)$_2$] in [Cu$_5$(CN)$_6$(DMF)$_4$] [60] or [Cu$_4$Zn(CN)$_6$(DMF)$_4$] [61], another topology, **rob**, is realized.

- Six-coordinated nodes with trigonal–prismatic coordination figure occur only in Eu[Ag(CN)$_2$]$_3$(H$_2$O)$_3$ where three terminal water molecules extend the Eu coordination number up to typical (CN = 9). The underlying topology is **acs**.
- Six-coordinated nodes with octahedral and trigonal–prismatic coordination figures occur only in the ratio 1:1 and in all cases give rise to the **nia** underlying net.

This list of correspondences can be extended with the nets with different degrees of nodes (e.g., four-, six-coordinated nets with tetrahedral and octahedral coordination figures in the ratio 1:2 result in the **fsh** topology), and the given data lead to similar conclusions as for two-periodic coordination polymers: the coordination figure strongly determines the underlying topology. Almost in all the cases, the shape of coordination polyhedron (considered without terminal ligands) coincides with the coordination figure of corresponding node in the most symmetrical embedding of the underlying net. The correspondences show that variation of the number and positions of the terminal ligands allows one to design the required underlying topology with a high probability. The cases with several possible underlying topologies (like **nom**, **nou**, **ptt**, or **uni**) have to be separately studied. Obviously, the shape, size, and number of terminal ligands as well as solvate molecules should be taken into account. The structure of the coordination center can also influence the underlying topology. Taking together all the factors could allow us to predict the 3D motifs more rigorously. We see that all these parameters sufficiently expand more general symmetry criteria. In more complicated structures containing different bridge ligands of different composition, the correlations could be more subtle, but they could be somehow catalogued. This is an important task of crystal design.

Similar regularities control crystal structures of other nature. Thus, Baburin and Blatov [23] showed that the topology of 3D frameworks of hydrogen-bonded molecules depends on the local arrangement of the molecule. They showed that strongly distorted tetrahedral environment of the molecule results in **qtz** or **dmp** topologies, while the square-like coordination leads to **cds** or **lvt** underlying topologies. Baburin [26] analyzed the influence of arrangement of active centers of hydrogen bonding over the molecule surface on the topology of molecular packings in coordination compounds. For instance, octahedral complexes with six ligands containing functional groups involved into hydrogen bonding prefer to form the **pcu** motif.

The underlying topology can be predetermined even more precisely if one extends the coordination figure to some larger part of the structure framework, that is, inherent to the resulting simplified net. This part can be referred to as a *secondary building unit* (SBU) of the net. To predict the net topology with a particular SBU, one has to know how specific is this SBU, and to what number of different nets it corresponds. For example, using the adamantane-like units one can construct the **dia** underlying net with much greater probability than starting from tetrahedral complexes. This approach lies in the base of reticular chemistry and was successfully used to design and synthesize new MOF materials [21, 30, 62].

In summary, the chemical factors influencing the local geometric and topological properties of the structure determine the underlying net topology according to the five-level scheme: *coordination number–complex group–coordination figure – secondary building unit–underlying topology* (Figure 1.12). If the ligands are either monodentate (terminal) or rather short bidentate-bridge, the scheme can be easily interpreted. The coordination number of the complexing atom or a cluster predetermines the maximum number of connected ligands in the complex group and, hence, the maximum number of nodes in the coordination figure; within this number, the coordination figure can be modified by introducing terminal ligands. The same coordination figure can correspond to different complex groups where the ligands vary by composition, shape, and size. At last, the same coordination figure can be built in different SBUs that provide different underlying topologies, thanks to various interstitial species or thermodynamic conditions [46, 47]. Thus, there is some diversity in final topology of the underlying net, but, as was shown above, this diversity is not so vast.

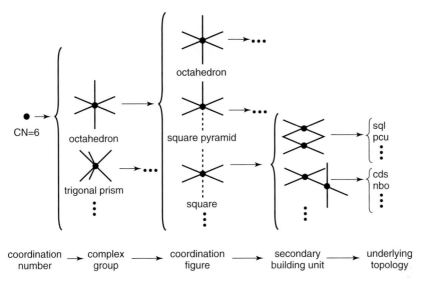

Figure 1.12 An example of the five-level hierarchical scheme of the underlying topology prediction starting from a complexing atom with coordination number 6, for example, Fe, Co, or Ni.

1.8
Outlook

The investigations of previous years mainly answered Wells' question – indeed, only rather small set of nets frequently occurs in crystal structures and we know almost all nets from this set. However, a number of new challenges emerged that will determine the development of periodic-graph approaches in the near future. We see the following main problems that should be solved at first: (i) it should become a standard for any structure investigation to provide the information on the underlying topology of possible structure representations along with crystallographic data; (ii) crystallographic databases should be integrated with the crystallochemical databases; (iii) the crystallochemical databases should be completed with all 2D and 3D representations of all known crystal structures that will enable the user to find similar topological motifs in different structures and to foresee possible topologies in new compounds by looking for chemical and structural prototypes in the databases; (iv) a database should be created that will collect strict interrelations between different topological representations like the interrelations "net–dual net," "net–edge net," "net–partial ring net," and so on. to let the user know in what structure representation he can get the required net topology; (v) the known parameters that influence the underlying topology (Section 1.7) should be deeper explored and new ones have to be discovered; (vi) an expert system should be developed that will use all the above-mentioned correlations and predict the structural dependencies in the form "complexing atom

+ ligands → list of possible complex groups → list of possible SBUs → list of underlying topologies" with occurrence probabilities and allowable space-group symmetries. Successful solution of these problems will inspire further progress in materials science and crystal chemistry.

References

1. Öhrström, L. and Larsson, K. (2005) *Molecule-based Materials: The Structural Network Approach*, Elsevier, Amsterdam.
2. Carlucci, L., Ciani, G., and Proserpio, D.M. (2007) in *Making Crystals by Design – Methods Techniques and Applications* (eds D. Braga and G. Grepioni), Wiley-VCH Verlag GmbH, Weinheim, pp. 58–85.
3. Wells, A.F. (1977) *Three-Dimensional Nets and Polyhedra*, Interscience, New York.
4. Chung, S.J., Hahn, Th., and Klee, W.E. (1984) Nomenclature and generation of three-periodic nets: the vector method. *Acta Cryst.*, **A40**, 42–50.
5. Klein, H.-J. (1996) Systematic generation of models for crystal structures. *Math. Model. Sci. Comput.*, **6**, 325–330.
6. Delgado-Friedrichs, O. and O'Keeffe, M. (2003) Identification of and symmetry computation for crystal nets. *Acta Cryst.*, **A59**, 351–360.
7. Klee, W.E. (2004) Crystallographic nets and their quotient graphs. *Cryst. Res. Technol.*, **39**, 959–968.
8. Eon, J.-G. (2005) Graph-theoretical characterization of periodicity in crystallographic nets and other infinite graphs. *Acta Cryst.*, **A61**, 501–511.
9. Thimm, G. (2009) Crystal topologies: the achievable and inevitable symmetries. *Acta Cryst.*, **A65**, 213–226.
10. Goetzke, K., Klein, J.-H., and Kandzia, P. (1988) in Graph-Theoretic Concepts in Computer Science, Proceedings of the International Workshop WG '87, June 29 – July 1, 1987, Kloster Banz/Staffelstein, FRG (eds H. Göttler and H.J. Schneider), Springer-Verlag, Berlin, pp. 242–254.
11. Koch, E., Fischer, W., and Sowa, H. (2006) Interpenetration of homogeneous sphere packings and of two-periodic layers of spheres. *Acta Cryst.*, **A62**, 152–167.
12. Delgado-Friedrichs, O., Foster, M.D., O'Keeffe, M., Proserpio, D.M., Treacy, M.M.J., and Yaghi, O.M. (2005) What do we know about three-periodic nets? *J. Solid State Chem.*, **178**, 2533–2554.
13. Brunner, G.O. and Laves, F. (1971) Zum problem der koordinationszahl. *Wiss. Z. Techn. Univ. Dres.*, **20**, 387–390.
14. Goetzke, K. and Klein, J.-H. (1991) Properties and efficient algorithmic determination of different classes of rings in finite and infinite polyhedral networks. *J. Non-Cryst. Solids*, **127**, 215–220.
15. O'Keeffe, M. and Hyde, S.T. (1997) Vertex symbols for zeolite nets. *Zeolites*, **19**, 370–374.
16. Blatov, V.A. (2007) Topological relations between three-dimensional periodic nets. I. Uninodal nets. *Acta Cryst.*, **A63**, 329–343.
17. Fischer, W. and Koch, E. (1989) Genera of minimal balance surfaces. *Acta Cryst.*, **A45**, 726–732.
18. Delgado-Friedrichs, O., O'Keeffe, M., and Yaghi, O.M. (2003) Three-periodic nets and tilings: regular and quasiregular nets. *Acta Cryst.*, **A59**, 22–27.
19. Hyde, S.T., Delgado Friedrichs, O., Ramsden, S.J., and Robins, V. (2006) Towards enumeration of crystalline frameworks: The 2D hyperbolic approach. *Solid State Sci.*, **8**, 740–752.
20. Blatov, V.A., Carlucci, L., Ciani, G., and Proserpio, D.M. (2004) Interpenetrating metal–organic and inorganic 3D networks: a computer-aided systematic investigation. Part I. Analysis of the Cambridge structural database. *CrystEngComm.*, **6**, 377–395.
21. Ockwig, N.W., Delgado-Friedrichs, O., O'Keeffe, M., and Yaghi, O.M. (2005) Reticular chemistry: occurrence and taxonomy of nets and grammar for the

design of frameworks. *Acc. Chem. Res.*, **38**, 176–182.
22. Baburin, I.A., Blatov, V.A., Carlucci, L., Ciani, G., and Proserpio, D.M. (2005) Interpenetrating metal–organic and inorganic 3D networks: a computer-aided systematic investigation. Part II. Analysis of the inorganic crystal structure database (ICSD). *J. Solid State Chem.*, **178**, 2452–2474.
23. Baburin, I.A. and Blatov, V.A. (2007) Three-dimensional hydrogen-bonded frameworks in organic crystals: a topological study. *Acta Cryst.*, **B63**, 791–802.
24. Baburin, I.A., Blatov, V.A., Carlucci, L., Ciani, G., and Proserpio, D.M. (2008) Interpenetrated 3D networks of H-bonded organic species: a systematic analysis of the Cambridge structural database. *Cryst. Growth Des.*, **8**, 519–539.
25. Baburin, I.A., Blatov, V.A., Carlucci, L., Ciani, G., and Proserpio, D.M. (2008) Interpenetrated three-dimensional hydrogen-bonded networks from metal–organic molecular and one- or two-dimensional polymeric motifs. *CrystEngComm.*, **10**, 1822–1838.
26. Baburin, I.A. (2008) Hydrogen-bonded frameworks in molecular metal–organic crystals: the network approach. *Z. Kristallogr.*, **223**, 371–381.
27. Blatov, V.A. (2006) Multipurpose crystallochemical analysis with the program package TOPOS. *IUCr Comp. Commun. Newsl.*, **7**, 4–38.
28. Blatov, V.A., Delgado-Friedrichs, O., O'Keeffe, M., and Proserpio, D.M. (2007) Three-periodic nets and tilings: natural tilings for nets. *Acta Cryst.*, **A63**, 418–425.
29. O'Keeffe, M., Peskov, M.A., Ramsden, S.J., and Yaghi, O.M. (2008) The reticular chemistry structure resource (RCSR) database of, and symbols for, crystal nets. *Acc. Chem. Res.*, **41**, 1782–1789.
30. Yaghi, O.M., O'Keeffe, M., Ockwig, N.W., Chae, H.K., Eddaoudi, M., and Kim, J. (2003) Reticular synthesis and the design of new materials. *Nature*, **423**, 705–714.
31. Delgado-Friedrichs, O. and O'Keeffe, M. (2005) Crystal nets as graphs: terminology and definitions. *J. Solid State Chem.*, **178**, 2480–2485.
32. Francl, M. (2009) Stretching topology. *Nat. Chem.*, **1**, 334–335.
33. Proserpio, D.M. (2010) Polycatenation weaves a 3D web. *Nat. Chem.*, **2**, 435–436.
34. Koch, E. and Fischer, W. (1978) Types of sphere packings for crystallographic point groups, rod groups and layer groups. *Z. Kristallogr.*, **148**, 107–152.
35. Thimm, G. and Klee, W.E. (1997) Zeolite cycle sequences. *Zeolites*, **19**, 422–424.
36. O'Keeffe, M. and Hyde, B.G. (1996) *Crystal Structures. I. Patterns and Symmetry*, Mineralogical Society of America, Washington, DC.
37. Delgado-Friedrichs, O., O'Keeffe, M., and Yaghi, O.M. (2007) Taxonomy of periodic nets and the design of materials. *Phys. Chem. Chem. Phys.*, **9**, 1035–1043.
38. Bonneau, C., Delgado Friedrichs, O., O'Keeffe, M., and Yaghi, O.M. (2004) Three-periodic nets and tilings: minimal nets. *Acta Cryst.*, **A60**, 517–520.
39. Sowa, H., Koch, E., and Fischer, W. (2007) Orthorhombic sphere packings. II. Bivariant lattice complexes. *Acta Cryst.*, **A63**, 354–364.
40. Sowa, H. (2009) Three new types of interpenetrating sphere packings. *Acta Cryst.*, **A65**, 326–327.
41. Blatov, V.A. and Proserpio, D.M. (2009) Topological relations between three-periodic nets. II. Binodal nets. *Acta Cryst.*, **A65**, 202–212.
42. Ramsden, S.J., Robins, V., and Hyde, S.T. (2009) Three-dimensional Euclidean nets from two-dimensional hyperbolic tilings: kaleidoscopic examples. *Acta Cryst.*, **A65**, 81–108.
43. Blatov, V.A. (2006) A method for hierarchical comparative analysis of crystal structures. *Acta Cryst.*, **A62**, 356–364.
44. O'Keeffe, M., Eddaoudi, M., Li, H., Reineke, T., and Yaghi, O.M. (2000) Frameworks for extended solids: geometrical design principles. *J. Solid State Chem.*, **152**, 3–20.
45. Dolomanov, O.V., Bourhis, L.J., Gildea, R.J., Howard, J.A.K., and Puschmann,

H. (2009) *OLEX2*: a complete structure solution, refinement and analysis program. *J. Appl. Cryst.*, **42**, 339–341.
46. Baburin, I.A., Leoni, S., and Seifert, G. (2008) Enumeration of not-yet-synthesized zeolitic zinc imidazolate MOF networks: a topological and DFT approach. *J. Phys. Chem. B*, **112**, 9437–9443.
47. Lewis, D.W., Ruiz-Salvador, A.R., Gómez, A., Rodriguez-Albelo, L.M., Coudert, F.-X., Slater, B., Cheetham, A.K., and Mellot-Draznieks, C. (2009) Zeolitic imidazole frameworks: structural and energetics trends compared with their zeolite analogues. *CrystEngComm.*, **11**, 2272–2276.
48. Blatov, V.A. and Peskov, M.V. (2006) A comparative crystallochemical analysis of binary compounds and simple anhydrous salts containing pyramidal anions LO_3 (L=S, Se, Te, Cl, Br, I). *Acta Cryst.*, **B62**, 457–466.
49. Klee, W.E. (1974) Al/Si distributions in tectosilicates: a graph-theoretical approach. *Z. Kristallogr.*, **140**, 154–162.
50. Fischer, W. and Koch, E. (1978) Limiting forms and comprehensive complexes for crystallographic point groups, rod groups and layer groups. *Z. Kristallogr.*, **147**, 255–273.
51. Delgado-Friedrichs, O., O'Keeffe, M., and Yaghi, O.M. (2006) Three-periodic nets and tilings: edge-transitive binodal structures. *Acta Cryst.*, **A62**, 350–355.
52. Delgado-Friedrichs, O. and O'Keeffe, M. (2007) Three-periodic tilings and nets: face-transitive tilings and edge-transitive nets revisited. *Acta Cryst.*, **A63**, 344–347.
53. Anurova, N.A. and Blatov, V.A. (2009) Analysis of ion-migration paths in inorganic frameworks by means of tilings and Voronoi–Dirichlet partition: a comparison. *Acta Cryst.*, **B65**, 426–434.
54. O'Keeffe, M. (1992) Self-dual plane nets in crystal chemistry. *Aust. J. Chem.*, **45**, 1489–1498.
55. Virovets, A.V., Blatov, V.A., and Peresypkina, E.V., unpublished results.
56. Yuan, A.-H., Zhou, H., Chen, Y.-Y., and Shen, X.-P. (2007) A novel three-dimensional Cu(II)–Mo(IV) bimetallic complex: synthesis, crystal structure, and magnetic properties. *J. Mol. Struct.*, **826**, 165–169.
57. Schwarz, P., Eller, S., Siebel, E., Soliman, T.M., Fischer, R.D., Apperley, D.C., Davies, N.A., and Harris, R.K. (1996) Template-driven syntheses of polymeric metal cyanides: a chiral nanoporous host for the $n\text{Bu}_4\text{N}^+$ ion. *Angew. Chem. Int. Ed.*, **35**, 1525–1527.
58. Lefebvre, J., Chartrand, D., and Leznoff, D.B. (2007) Synthesis, structure and magnetic properties of 2-D and 3-D [cation]{M[Au(CN)$_2$]$_3$}(M =Ni, Co) coordination polymers. *Polyhedron*, **26**, 2189–2199.
59. Niu, T., Lu, J., Wang, X., Korp, J.D., and Jacobson, A.J. (1998) Syntheses and structural characterizations of two three-dimensional polymers: [{(n-C$_4$H$_9$)$_3$Sn}$_3$M(CN)$_6$](M=Fe, Co). *Inorg. Chem.*, **37**, 5324–5328.
60. Peng, S.-M. and Liaw, D.-S. (1986) Cu(II) ion catalytic oxidation of o-phenylenediamine and diaminomaleonitrile and the crystal structure of the final products (C$_{12}$N$_4$H$_{11}$)(ClO$_4$) H$_2$O and [Cu$_5$(CN)$_6$(dmf)$_4$]. *Inorg. Chim. Acta*, **113**, L11–L12.
61. Cui, C.-P., Lin, P., Du, W.-X., Wu, L.-M., Fu, Z.-Y., Dai, J.-C., Hu, S.-M., and Wu, X.-T. (2001) Synthesis and structures of two novel heterometallic polymers. *Inorg. Chem. Commun.*, **4**, 444–446.
62. Eddaoudi, M., Moler, D.B., Li, H., Chen, B., Reineke, T.M., O'Keeffe, M., and Yaghi, O.M. (2001) Modular chemistry: secondary building units as a basis for the design of highly porous and robust metal-organic carboxylate frameworks. *Acc. Chem. Res.*, **34**, 319–330.

2
Energy Landscapes and Structure Prediction Using Basin-Hopping
David J. Wales

2.1
Introduction

The potential energy surface (PES) or potential energy landscape [1] governs the structure, dynamics, and thermodynamics of any system within the Born–Oppenheimer approximation [2] for a particular electronic state. Breakdown of the Born–Oppenheimer approximation is important for understanding effects such as nonadiabatic transitions, but in the present chapter we will restrict attention to a single PES, and neglect effects such as surface crossings [3, 4]. Whilst the simplest Monte Carlo (MC) calculations require only a value for the potential energy, molecular dynamics (MD) simulations need both the energy and gradient [5, 6]. Today the MC and MD approaches are still the principal workhorses used in computer simulation of atomic, molecular, and condensed matter systems. However, a complementary class of simulation methods exists, where geometry optimization techniques are employed to survey the PES, usually through the characterization of stationary points, where the gradient of the potential vanishes. A coarse-grained description involving local minima, and the pathways that connect them via transition states, can be used to predict global thermodynamics and kinetics, as well as structure. Here we define a transition state according to the geometrical criterion of Murrell and Laidler, as a stationary point with a single imaginary normal mode frequency [7]. The information required for the efficient geometry optimization is essentially the same as for molecular dynamics, i.e., the potential energy and the gradient for any given configuration. Basin-hopping global optimization [1, 8–10], which is the main focus of this chapter, involves successive minimization, as discussed in Section 3.

Global thermodynamic properties can be calculated from a database of local minima using the superposition approach [1, 11–18], where the total partition function is written as a sum of contributions from different structures. However, a collection of local minima does not in itself constitute an energy landscape, because there is no measure of connectivity, and hence no way to address kinetics. The interplay between kinetics and thermodynamics actually plays a key role in defining the "structure-seeking" properties of a given system, as discussed in Section 2.

Modern Methods of Crystal Structure Prediction. Edited by Artem R. Oganov
Copyright © 2011 WILEY-VCH Verlag GmbH & Co. KGaA, Weinheim
ISBN: 978-3-527-40939-6

To construct a kinetic transition network [19–26] based upon stationary points of the PES requires additional calculations of transition states and the approximate steepest-descent paths that define the connectivity between pairs of local minima. Characterizing transition states using tools such as the doubly-nudged [27] elastic band [28–30] method, followed by hybrid eigenvector-following [31, 32] for accurate refinement, enables both high- and low-barrier transition mechanisms to be identified. When combined with standard unimolecular rate theory [33–35] to define minimum-to-minimum rate constants, this approach provides access to time and length scales beyond the reach of conventional simulations. Networks designed to represent the kinetic properties of interest can be expanded systematically using the discrete path sampling approach [1, 20, 36], which has now been applied to a wide variety of atomic and molecular systems [37–39].

Further details of the geometry optimization algorithms used to locate stationary points can be found in the above references. Details of the interatomic or intermolecular potentials used to calculate the energy and gradient, which extends to explicit treatment of the electronic structure in some cases, will also be omitted. Instead we will focus on generic features of the energy landscape that affect structure prediction (Section 2), and on basin-hopping global optimization [1, 8, 10, 40] (Section 3). This approach is probably unrivalled in terms of the diversity of applications treated over the last decade, and some illustrative examples will be given in Section 3, with particular emphasis on condensed matter systems in Section 4. The time taken to locate the global minimum is governed by the topology of the underlying energy landscape, and this connection is illustrated using disconnectivity graphs in Section 2. Visualizing the landscape in this way has provided fundamental insight into the connections between structure prediction and global thermodynamic and kinetic properties [1]. The first results of this kind for condensed matter systems are summarized in Section 4, and reveal a striking contrast between the potential energy surfaces of systems that crystallize readily and those that are prone to form glasses. This organization makes prediction of the underlying crystal significantly harder for good glass-formers.

2.2
Visualizing the Landscape

The capability to visualize potential energy surfaces in a meaningful way has played a key role in recent efforts to understand how observable properties such as structure, dynamics, and thermodynamics are encoded in the landscape [1, 41, 42]. The disconnectivity graph approach discussed below has features in common with the "energy lid" and "energy threshold" procedures of Sibani, Schön, and coworkers [43–46]. Disconnectivity graphs can be constructed using definitions of connectivity based on potential energy, free energy, or interconversion rates [1, 41, 47–50], and avoid the "filling-in" problem [26, 51–54] associated with free energy surfaces projected onto one or two degrees of freedom. While projection may lead to overlap of conformations that are actually far apart in configuration space [26], the disconnectivity graph construction preserves the corresponding barriers.

Some simple one-dimensional functions and corresponding disconnectivity graphs are shown in Figure 2.1. The graph is constructed by dividing the local minima into disjoint sets, or "superbasins" [48], for a series of threshold potential energies, $V_1 < V_2 < V_3 < \ldots$. For a threshold V_α, two minima lie in the same superbasin if they can interconvert via one or more transition states without exceeding the threshold. Each superbasin is represented by a node (point) on the horizontal axis, while the vertical axis represents increasing energy. The nodes for successive energy thresholds are joined if they share common minima, and each branch starts at low energy from the value corresponding to a particular local minimum. At low energy the graph consists of a single branch that emanates from the global minimum, while at very high energy there is again a single branch

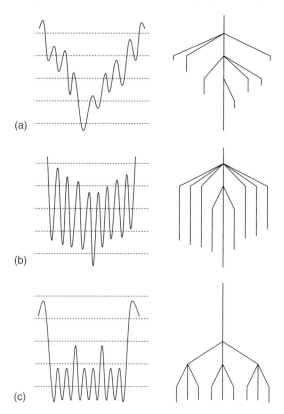

Figure 2.1 One-dimensional potential energy functions (left) and the disconnectivity graphs that they correspond to (right). The energies at which the superbasin analysis was performed are indicated by dotted blue lines. (a) "Palm tree": low downhill barriers and a well-defined global minimum. (b) "Willow tree": a well-defined global minimum, but higher downhill barriers. (c) "Banyan tree": no well-defined global minimum and barrier heights larger than the typical energy difference between successive minima.

corresponding to all the minima. At intervening energies, branches corresponding to new local minima appear, and merge into a superbasin at the first V_α threshold that exceeds the lowest energy transition state they are connected to. If the spacings between the V_α were infinitesimally small, then every transition state would give rise to a bifurcation of branches, which can obscure the features of interest. Too large a spacing can coarse-grain over too much of this structure, but it is generally straightforward to choose a value that highlights the key changes in connectivity. The arrangement of the nodes on the horizontal axis is usually chosen to provide the clearest picture of the landscape, although other representations are possible [55]. Such choices have no bearing on quantitative calculations of global thermodynamics and kinetics, where appropriate densities of states are used for each stationary point [1, 42]. However, the disconnectivity graph visualization can serve as a "zeroth-order" calculation, which enables us to predict the appearance of heat capacity peaks and distinct relaxation time scales corresponding to different morphologies, as discussed below.

Free energy disconnectivity graphs can be constructed using the mincut–maxflow algorithm to define the free energies of transition states [19, 53], or by a recursive regrouping scheme where the free energy of group J is defined as

$$F_J(T) = -k_B T \ln \sum_{j \in J} Z_j(T) \tag{2.1}$$

where $Z_j(T)$ is the canonical partition function for minimum j. The regrouping of states can be achieved by merging free energy minima separated by free energy barriers below a threshold $\Delta F_{\text{barrier}}$ [37, 56]. The free energy of the group of transition states that directly connect minima in group J to minima in group L is defined as

$$F_{LJ}^\dagger(T) = -k_B T \ln \sum_{l \leftarrow j} Z_{lj}^\dagger(T) \equiv -k_B T \ln Z_{LJ}^\dagger(T) \tag{2.2}$$

and the intergroup rate constant from J to L, k_{LJ}, is then [37, 57]

$$k_{LJ}(T) = \sum_{l \leftarrow j} \frac{p_j^{\text{eq}}(T)}{p_J^{\text{eq}}(T)} k_{lj}(T) = \sum_{l \leftarrow j} \frac{Z_j(T)}{Z_J(T)} \frac{k_B T}{h} \frac{Z_{lj}^\dagger(T)}{Z_j(T)}$$

$$= \frac{k_B T}{h} \frac{Z_{LJ}^\dagger(T)}{Z_J(T)} = \frac{k_B T}{h} e^{-[F_{LJ}^\dagger(T) - F_J(T)]/k_B T} \tag{2.3}$$

where $Z_{lj}^\dagger(T)$ is the partition function of transition state $\dagger$ connecting minimum j to minimum l, with the coordinate corresponding to the unique negative Hessian eigenvalue excluded, and $k_{lj}(T) = k_B T Z_{lj}^\dagger(T)/h Z_j(T)$ is the transition state theory [33] expression for the corresponding rate constant. In the calculations presented below, harmonic densities of states are employed to calculate the vibrational partition functions, but anharmonic and quantum mechanical corrections can also be used [58–60].

Three different patterns of organization are revealed in the disconnectivity graphs for the model landscapes in Figure 2.1. Since cutting any edge separates the structure into two parts, each one is a tree graph [69] and appropriate names have previously been suggested for each pattern [49]. The landscape in Figure 2.1a, with a well-defined global minimum, and relatively small downhill barriers, reminds us of a "palm tree," while the larger barriers in Figure 2.1b produce a "weeping willow." The topology in Figure 2.1c is qualitatively different, since there are distinct energy scales for the barriers, which lead to a hierarchical structure. The resulting pattern is reminiscent of a "banyan tree," where there are numerous branches leading to low-lying minima that are separated from the global minimum by high barriers. These competing minima can be connected to quantitative measures of "frustration" in the energy landscape [70, 71]; an extreme example corresponding to a glassy landscape will be discussed in Section 4.

Some specific examples are shown in Figure 2.2. The palm tree motif is illustrated by parts (a)–(d), for (a) an atomic cluster with a Mackay icosahedron [72] as the global minimum, (b) a self-assembling icosahedral shell [41], (c) a supercell representation of bulk silicon [64], and (d) a polyalanine peptide. The willow tree pattern revealed for C_{60} in Figure 2.2e is the result of a well-defined global minimum, but large barriers for interconverting isomers with different σ-bonding frameworks. Finally, the graph for $(H_2O)_{20}$ in Figure 2.2f has a hierarchical structure, where sets of minima are disconnected together when certain edges of the graph are removed. The form of these graphs provides a very convenient way to think about the likely efficiency of relaxation to the global minimum. A good "structure-seeking" system needs to possess a well-defined free energy minimum, which is kinetically accessible over the temperature range of interest. The interplay of thermodynamic and kinetic factors is particularly noteworthy, and cannot be described without a measure of connectivity between the local minima. It is the palm tree graph that we associate with structure-seeking properties. However, the willow tree graph also has a well-defined global minimum, and efficient relaxation is again possible for a sufficiently high temperature, in good agreement with experimental routes to buckminsterfullerene [49, 67, 73].

The identification of structure-seeking, self-assembling, or self-organizing systems with the palm tree pattern is consistent with the "folding funnel" view of protein folding, described in terms of a set of kinetically convergent pathways [74]. A well-defined free energy minimum exists for such landscapes over a range of temperature [42], and the lack of competitive structures can be identified with "minimal frustration" [75].

Before comparing glassy and crystalline landscapes in Section 4 it is instructive to consider a double-funnel landscape, which contains two palm tree features. Here we consider a cluster of 38 atoms bound by the LJ potential [61], which has served as a benchmark for global optimization [1, 8–10], thermodynamic sampling [76–79], and rare event dynamics [1, 20, 36]. The global potential energy minimum is a truncated octahedron, which is separated by a high barrier from structures based on incomplete Mackay icosahedra [77]. The octahedral region of configuration space is favored by potential energy, while the icosahedral region is favored by entropy, and the change in the global free energy minimum with increasing temperature provides a finite system analog of a first order phase transition. This transition

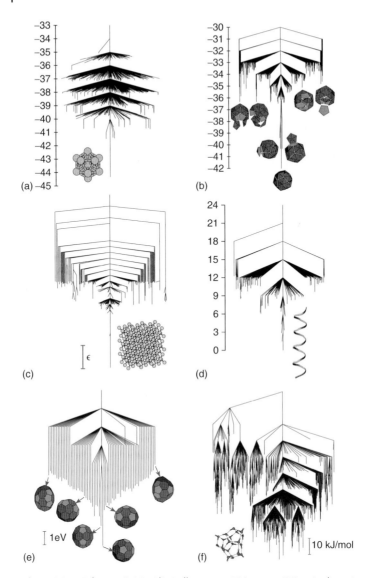

Figure 2.2 "Palm tree" (a)–(d) "willow tree" (e), and "banyan tree" (f) disconnectivity graphs for the following systems: (a) a cluster of 13 atoms bound by the Lennard–Jones (LJ) potential [61], LJ_{13}, including 1467 distinct local minima [62]. The energy is in units of the pair well depth. (b) An icosahedral shell composed of 12 pentagonal pyramids [41]. (c) A bulk representation of silicon using the Stillinger–Weber potential [63] and a supercell containing 216 atoms [64]. ϵ is the pair well depth. (d) The polyalanine peptide ala_{16} represented by the AMBER95 potential [65] and a distance-dependent dielectric [66]. The energy is in kcal/mol relative to the global minimum. (e) C_{60} using a density functional theory treatment of the electronic structure [67]. (f) An $(H_2O)_{20}$ cluster bound by the TIP4P [68] potential [49]. (Please find a color version of this figure on the color plates.)

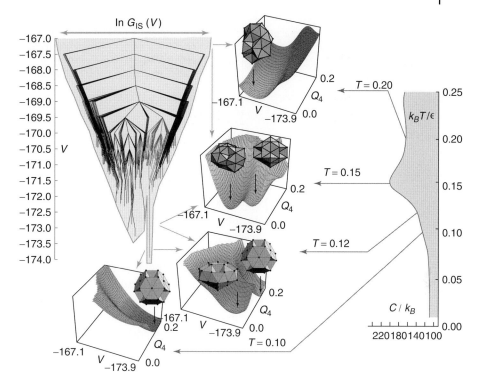

Figure 2.3 disconnectivity graph, heat capacity, and free energy surfaces for the double funnel LJ$_{38}$ cluster [42]. Here the width of the disconnectivity graph has been scaled according to the natural logarithm of the total number of minima with potential energy less than V, $\ln G_{IS}(V)$. A representative structure is illustrated above each free energy minimum. (Please find a color version of this figure on the color plates.)

gives rise to a feature in the heat capacity at around $k_B T/\epsilon = 0.12$, corresponding to a solid–solid transformation (Figure 2.3). Here, ϵ is the pair well depth for the LJ potential, and the broad peak at higher temperature corresponds to the melting transition. Figure 2.3 also includes a scaled disconnectivity graph and free energy surfaces projected onto the potential energy, V, and a bond-orientational order parameter, Q_4 [80, 81]. At low temperature there is a single free energy minimum corresponding to the fcc configurations, but a double minimum with a barrier along the Q_4 axis emerges at $k_B T/\epsilon = 0.12$, where fcc and icosahedral structures are equally populated. Around the melting temperature the free energy surface exhibits a double minimum along the V axis, corresponding to contributions from solid-like and liquid-like regions of configuration space. Finally, at high temperature there is again a single free energy minimum, this time corresponding to the liquid-like state [42].

The first lesson to emerge from this study is that competing morphologies corresponding to separate potential energy funnels are likely to produce features in the heat capacity. The double funnel structure also results in a separation

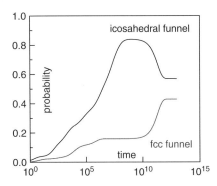

Figure 2.4 Occupation probabilities of local minima based on fcc and icosahedral packing starting from an equilibrium high temperature distribution [82]. Time is in reduced units of $\sqrt{m\sigma^2/\epsilon}$ for the LJ potential.

of relaxation time scales [1, 82], as shown in Figure 2.4. The fast component corresponds to direct relaxation to the fcc region, while the slow component corresponds to the probability of reaching a low-lying icosahedral structure first, followed by escape to the fcc region to achieve true equilibrium corresponding to a time scale of around 10^{12} reduced time units. The nonequilibrium distribution corresponding to local relaxation persists for about four decades in time [82].

These observations are directly relevant for structure prediction. Global optimization algorithms based on annealing are very inefficient for this system [83] and the performance will be even worse for more complicated landscapes. Trapping in a metastable free energy minimum that lies behind an insurmountable barrier is the likely outcome of annealing on a glassy landscape. Alternative approaches, such as basin-hopping [1, 8–10, 40], are needed to tackle such landscapes efficiently.

2.3
Basin-Hopping Global Optimization

The general principle that underpins basin-hopping global optimization is to propose moves that change a current structure, minimise from this starting point, and accept or reject the local minimum that results based upon criteria such as the change in energy. If the step is accepted then the local minimum becomes the starting point for the next perturbation of the geometry. The lowest minima encountered during the run are the structures that are most likely to be favorable at temperatures corresponding to a solid-like phase. Because the energy is minimized after the proposed step the perturbations can be much larger than the displacements used in typical thermodynamic sampling schemes. Ideally, we wish to take steps that move the system into the catchment basin of a neighboring local minimum, without jumping too far through configuration space and skipping over adjacent structures.

Basin-hopping [1, 8–10] is a generalization of the "Monte Carlo plus energy minimization" procedure of Li and Scheraga [40], where the coordinates are not necessarily reset to those of the current minimum. In fact, resetting is usually found to be the most effective strategy [84], but it is the minimization that is crucial for efficient global optimization. A survey revealed that most studies reporting new global minima for LJ clusters had actually applied a minimization step [8, 85–88]. For example, Deaven and Ho described the first genetic algorithm [89–93] study to include minimization of the population of structures after each step [87, 94]. The minimization has been interpreted as a Lamarckian rather than Darwinian evolution [95], where parents pass on features that they have acquired, rather than inherited. The transformed potential energy for a $3N$-dimensional configuration $\mathbf{X}$, where N is the number of atoms, is

$$\widetilde{V}(\mathbf{X}) = \min\{V(\mathbf{X})\} \qquad (2.4)$$

where $V(\mathbf{X})$ is the original potential energy, and "min" signifies that energy minimization is carried out from $\mathbf{X}$. The potential energy landscape is therefore transformed into the set of catchment basins for the local minima, as shown in Figures 2.5 and 2.6. All the results described below employed the LBFGS minimization algorithm of Liu and Nocedal [96]. A modified version of this routine, with line searches removed, is coded in the GMIN [97] and OPTIM [98] programs, which are available for global optimization and general geometry optimization under the Gnu public license.

If the catchment basins are defined by steepest-descent paths according to a first-order differential equation then the boundaries cannot interpenetrate [1]. More complex catchment areas with interpenetrating regions are observed when efficient energy minimization schemes are employed [99, 100], such as the LBFGS method [96]. The basin transformation corresponding to the minimization step must be combined with a step-taking scheme, and a number of different approaches are available in the GMIN program. The simplest approach, which is generally quite effective, is to perturb all the Cartesian coordinates using displacements drawn randomly from a uniform distribution in the range $\max_d \times [-1, 1]$, where $\max_d$ is a parameter that specifies the maximum displacement. A simple Monte Carlo-type sampling based on the energies of the previous and new local minima, V_{old} and

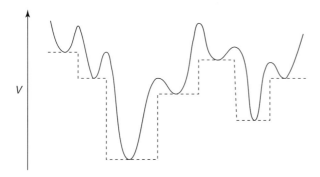

Figure 2.5 The basin transformation defined in Eq. (2.4) transforms the one-dimensional potential function, V, shown in black, to the value for the local minimum, $\widetilde{V}$, corresponding to a catchment basin (dashed lines).

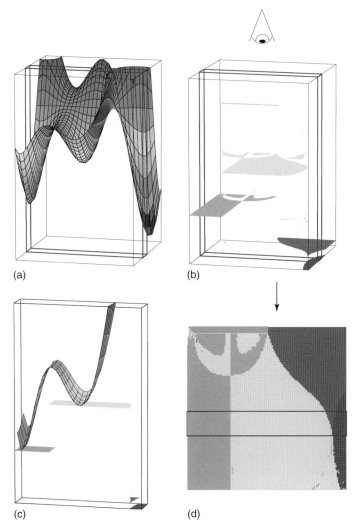

Figure 2.6 An alternative representation of the basin transformation for a two-dimensional potential function [10]. (a) Original landscape. (b) Transformed landscape. Every local minimum of $V(\mathbf{X})$ corresponds to a catchment basin of $\tilde{V}(\mathbf{X})$. The surfaces are colored according to the energy as for part (a). (c) Cut through the combined $\tilde{V}(\mathbf{X})$ and $V(\mathbf{X})$ surfaces for the boxed region highlighted in red in all the panels. (d) Top view of the transformed surface.

V_{new}, can then be applied. The step is accepted if $V_{\text{new}} < V_{\text{old}}$, or if $V_{\text{new}} > V_{\text{old}}$, and $\exp\{(V_{\text{old}} - V_{\text{new}})/k_B T\}$ is greater than a random number drawn from the range [0, 1]. Similar results are obtained for LJ clusters if threshold acceptance is used, where a move is accepted if $V_{\text{new}} - V_{\text{old}}$ lies below a certain cut-off [101]. An improvement in efficiency is obtained if the coordinates are reset to those of the latest minimum in the chain before the next step is proposed [84].

A particularly attractive feature of the above scheme is that the temperature used in the accept/reject step is the only variable parameter if the maximum step size, $\max_d$, is dynamically adjusted to give a fixed average acceptance ratio [8]. Hence it is usually straightforward to apply basin-hopping to a new system, without requiring a parameter optimization that is only possible once the global minimum is actually known.

Since minimization is the key step in basin-hopping it is possible to construct a reasonably successful algorithm using a variety of alternative step-taking schemes. A number of different methods have been proposed, ranging from lattice-based searches to moves that employ molecular dynamics [102–115]. Genetic algorithms that include minimization of the population are effectively searching the same transformed landscape, and should therefore give similar results. To understand why the minimization step is so important requires an analysis of the thermodynamics for the transformed landscape, as discussed below.

The LJ_{38} cluster has provided a useful test for global optimization algorithms since the fcc global minimum was first characterized. Starting from random distributions of atoms in a sphere, the mean first encounter time (MFET) for the simplest basin-hopping algorithm with GMIN was reported as around 2000 quenches (minimizations) in 1999 [10]. For any particular system, it is usually possible to improve the MFET by at least a factor of two through optimizing parameters such as the temperature, or the particular moves [1, 113, 116]. The current best effort for LJ_{38} with the GMIN program provides almost an order of magnitude improvement over the old result, and uses a combination of reseeding when no energy decrease in the lowest minimum has been achieved in a given number of cycles, a cyclic taboo list [117, 118] of structures that correspond to the lowest minima found at reseeding, and symmetrizing steps. The symmetrizing moves introduce some bias, but are justified from the principle of maximum symmetry, which states that structures with a higher (possibly continuous [119, 120]) symmetry measure are more likely to have particularly high or particularly low energies [1, 121, 122]. The MFET for LJ_{38} is then 204 quenches, which corresponds to 25,000 energy and gradient evaluations and 0.5 s of CPU time on a single Intel T9300 CPU (as found in the author's laptop).

Systems composed of distinct molecules pose additional problems for global optimization. For example, the hierarchical structure of the disconnectivity graph for $(H_2O)_{20}$ (Figure 2.2f) makes this a much more difficult target than an LJ cluster with a comparable number of degrees of freedom. Here, the problem is that different hydrogen-bonding patterns span a remarkably wide range of potential energy for a given arrangement of oxygen atoms [123–125]. The problems of crystal structure prediction for molecules with increasing hydrogen-bond donor–acceptor possibilities may well be due to analogous features of the energy landscape [126]. A significant improvement in efficiency for both neutral and protonated water clusters [123, 127–129] was gained by alternating blocks of translational moves and blocks of rotational moves. Some representative structures are shown in Figure 2.7. Allowing larger moves by setting a target acceptance ratio of 0.3 was also beneficial. Subsequent studies [130] have confirmed the original basin-hopping results for

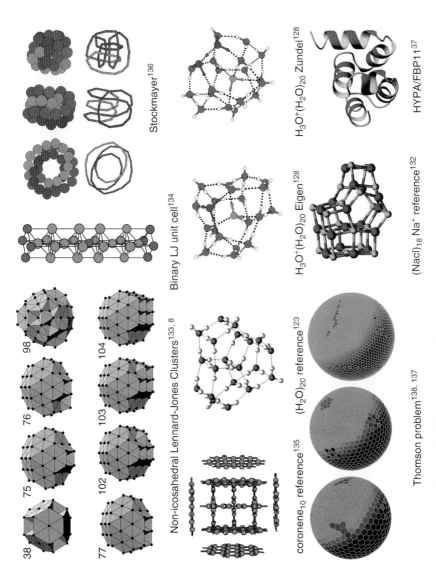

Figure 2.7 A selection of global minima from the Cambridge Cluster database at http://www.wales.ch.cam.ac.uk/CCD.html. (Please find a color version of this figure on the color plates.)

the TIP4P potential, and extended the range for which reliable global minima are available using graph theoretical analysis to locate favorable hydrogen-bonding patterns [131].

Clusters of polycyclic aromatic hydrocarbons, such as pyrene and coronene, pose problems because of their anisotropy. Nevertheless, basin-hopping has been successfully applied to such systems, using moves that allow both local and collective reorganization [135]. An example is illustrated in Figure 2.7, along with a number of other structures taken from the Cambridge Cluster Database (CCD) [139], including structure prediction for an FF domain of the 71 residue protein HYPA/FBP11 [37] (protein data bank code 1UZC) and an alkali halide cluster [132]. A wide variety of structures are available for download from the CCD, including results for both generic potentials and specific empirical potentials for transition metals [140, 141], fullerenes [142], and doped inert gas clusters [143]. Recent predictions (Figure 2.7) include remarkable knotted topologies for clusters bound by the Stockmayer potential (LJ plus points dipole) [136], and the evolution of defect structure for the Thomson problem [144] of N charges on a sphere [137, 138]. Although J. J. Thomson originally proposed the latter model to investigate atomic structure, the structures involved have been used to analyze "spherical crystallography" [145]. The simplest way to satisfy Euler's theorem for a spherical topology, and produce a disclination charge of 12, is for 12 particles to be five-coordinate. However, a wide variety of alternative defect structures arises for larger systems, where minimizing strain energy becomes important [146, 147]. Applications include fullerene structures [148, 149], spheroidal viruses [150, 151], colloidal silica microspheres [152], superconducting films [153, 154], lipid rafts deposited on vesicles [155], micropatterning of spherical particles, [156] "colloidosomes" [145, 157, 158], cell surface layers in prokaryotes [159, 160], and multielectron bubbles in superfluid helium [161, 162]. It will certainly be interesting to see whether the basin-hopping approach can provide similar improvements in crystal structure prediction for periodic systems in the future [163–169].

To conclude this section it is instructive to consider why the basin-hopping approach has been so successful for such a wide range of systems. The local minimization step clearly facilitates transitions between different minima, since the transition state regions are eliminated, and atoms or polymer chains can pass through one-another without encountering a barrier. However, the success of basin-hopping for multi-funnel potential energy landscapes requires a more detailed analysis, which focuses on the thermodynamics of the transformed landscape [9, 170]. The calculated equilibrium occupation probabilities as a function of temperature have little overlap for the original potential [9], so the chance of escaping from a low-lying icosahedral configuration to an fcc structure is low (Figure 2.8). The occupation probability of a local minimum for the original landscape depends on the normal mode frequencies in the harmonic approximation. However, on transformed landscape the relevant catchment volume, A, corresponds to the configuration space from which minimization leads to the structure in question. The average normal mode frequencies for minima assigned to the fcc, icosahedral, and liquid-like regions of the landscape are in the ratio

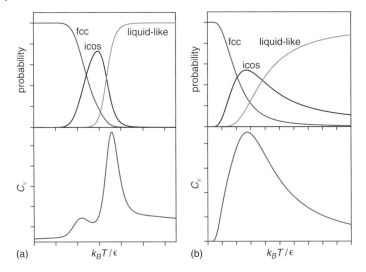

Figure 2.8 The occupation probability distributions for LJ$_{38}$ corresponding to the fcc, icosahedral, and "liquid-like" regions of configuration space. Overlap between the liquid-like and fcc regions extends over a much wider range of temperature for the transformed landscape (b) compared to the original potential (a) [9]. The corresponding heat capacities (bottom panels) reveal how the transitions are broadened for the transformed landscape [9].

$\bar{\nu}_{\text{fcc}} : \bar{\nu}_{\text{icos}} : \bar{\nu}_{\text{liquid}} = 1 : 0.968 : 0.864$. The occupation probabilities depend upon these ratios raised to the power $3N - 6$, and the stiffer frequencies of the fcc minima lead to a distinctly lower entropy [9]. In contrast, the catchment basin volumes follow the opposite trend [9], with $A_{\text{fcc}} : A_{\text{icos}} : A_{\text{liquid}} = 1 : 0.0488 : 0.00122$. Since the catchment volume appears in the numerator, rather than the denominator for the frequencies, the result is that changes in the global free energy minimum are significantly broader for the transformed landscape. A new global minimum therefore has a greater occupation probability at temperatures where the barrier between the two regions of configuration space is surmountable. Fortunately, there is no need to calculate any of these thermodynamic quantities in a basin-hopping run: the advantageous changes in occupation probabilities are obtained automatically via the minimization step.

2.4
Energy Landscapes for Crystals and Glasses

The energy landscape has been visualized for several crystalline systems modeled using periodic boundary conditions. The disconnectivity graphs in Figure 2.9 for bulk silicon, represented by a modified Stillinger–Weber (SW) silicon potential [63], and for a unit density LJ crystal, correspond to systems where basin-hopping can locate the global minimum in a relatively small number of cycles. The graph

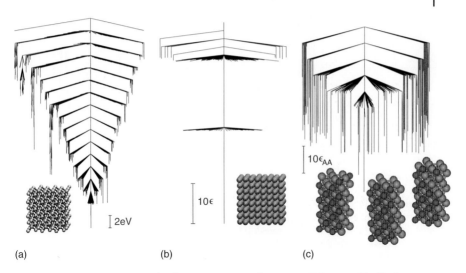

Figure 2.9 Disconnectivity graphs for crystalline systems with low-lying minima superimposed. (a) disconnectivity graph obtained at a constant pressure of one atmosphere for a modified Stillinger–Weber silicon potential where the magnitude of the three-body term is increased by 50%. The supercell contains 216 atoms [1]. (b) the lowest 500 minima in the vicinity of the unit density LJ crystal for a supercell containing 256 atoms [64]. (c) Disconnectivity graph for a binary LJ crystal [134] represented by a supercell containing 320 atoms in an A:B ratio of 80:20.

obtained for silicon at constant pressure in Figure 2.9 can be compared with the graph in Figure 2.2c, which corresponds to constant volume and an optimum box size for the crystal [64]. In contrast, the graph for the binary Lennard–Jones (BLJ) system (Figure 2.9) exhibits numerous relatively low-lying minima, even in the vicinity of the crystal [134]. BLJ models are popular for simulations of supercooled liquids and glasses [64, 173–181] because crystallization is not usually observed on the time scales accessible to MD. The BLJ mixture considered here is an 80:20 mixture of A:B atoms with parameters $\sigma_{AA} = 1$, $\sigma_{AB} = 0.8$, $\sigma_{BB} = 0.88$, $\epsilon_{AA} = 1$, $\epsilon_{AB} = 1.5$, and $\epsilon_{BB} = 0.5$ [173]. The units of distance and energy are chosen as σ_{AA} and ϵ_{AA}, and number densities between 1.1 and 1.3 are considered, using the minimum image convention [5]. The finite cutoff is usually dealt with by shifting the energy to be continuous at the cutoff, or shifting both the energy and gradient [182]. The latter scheme is used for all the calculations reported below.

The BLJ model was originally introduced [183] to represent the metallic glass $Ni_{0.8}P_{0.2}$. For suitable parameterizations the model certainly has glassy characteristics, and it has even been claimed that "This model is suitable for such problems because of the lack of a crystalline state." In fact, crystalline states have been identified, both for nonphase separated states using basin-hopping [134], and for an even lower energy phase separated state by construction [184]. A unit cell for the crystal that corresponds to space group $I4/mmm$, with a structure analogous to $Ir(UC)_2$, is shown in Figures 2.7 and 2.9.

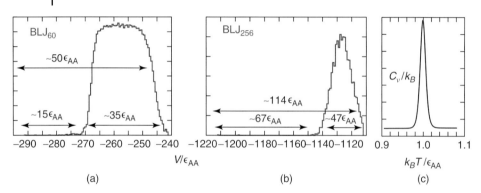

Figure 2.10 Probability distribution for local minima as a function of potential energy, V, in BLJ systems containing 60 (a) and 256 (b) atoms in the supercell [171]. (c) The heat capacity peak obtained for the melting transition with a supercell of 320 atoms at a number density of $1.2\sigma_{AA}^{-3}$.

Equilibrium thermodynamic properties have now been calculated for the above BLJ model using parallel tempering Monte Carlo methods [185, 186]. Simulations were conducted for supercells containing a total of 60, 256, and 320 atoms using 32 replicas [171], and the density of states was reconstructed for each system using a multiple histogram approach [187–189]. The problem of achieving a globally ergodic simulation is clear from the probability distribution for the local minima as a function of potential energy, V, in Figure 2.10. Comparing the distributions for supercells containing 60 and 256 atoms, we see that the separation between the lowest crystalline minimum and the peak corresponding to liquid-like minima scales extensively with system size, as it should. We were unable to obtain equilibrium sampling using a flat-histogram approach, but converged heat capacity curves were achieved using parallel tempering, with peaks corresponding to bulk melting at $k_B T/\epsilon_{AA} = 0.72$ and 0.99 for supercells containing 256 and 320 atoms, respectively [171], with a number density of $1.2\sigma_{AA}^{-3}$. Hence we can now be sure that previous simulations of supercooled BLJ systems conducted at lower temperatures really do correspond to the supercooled regime.

The fundamental importance of the understanding the glass transition is often emphasized [190, 191], and recent visualizations of the underlying PES provide a new perspective [171, 172, 192]. Figure 2.11 shows two alternative representations of the graph obtained by reconstructing all the transitions in a locally ergodic pathway [193], defined using an energy fluctuation metric [194]. The rearrangements defined by all the transition states in the stationary point database can be classified using a robust geometrical definition of a cage-breaking event [172]. Removing these transition states (Figure 2.11a) and coloring the separate portions of the graph according to the energy at which they are disconnected, reveals a higher level of organization in the landscape. Including cage-breaking rearrangements allows the system to explore virtually the whole landscape, but

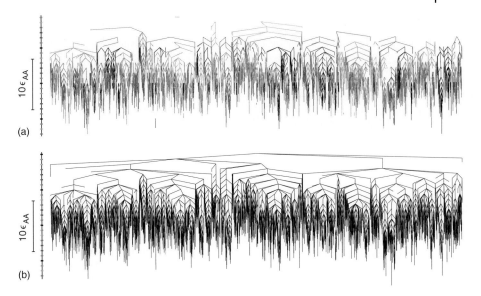

Figure 2.11 Two alternative representations of a disconnectivity graph obtained for a BLJ system with 60 atoms in the supercell at a number density of $1.3\sigma_{AA}^{-3}$ and $k_BT/\epsilon_{AA} = 0.713$. In (a), transition states corresponding to cage-breaking rearrangements are removed, while in (b) all the other transition states are omitted [172]. The graphs are colored according to the energy at which connection from the rest of the graph is lost, with a key on the vertical axis. The structure that results in (a) shows that the system can only explore local regions of configuration space when cage-breaking rearrangements are forbidden, leading to a higher order organization of the landscape. (Please find a color version of this figure on the color plates.)

removing them separates the landscape into distinct regions, each one containing a small number of deep potential energy funnels. This separation provides a possible definition of the "metabasin" structure suggested by previous work [195, 196].

The structure represented in Figure 2.11 provides a clear conceptual basis for the glassy behavior of this system. The potential energy landscape contains an exponentially large number of relatively low-lying local minima corresponding to disordered structures, separated by barriers that are very large compared to the available thermal energy of the supercooled liquid. As the system is progressively supercooled an increasing proportion of the landscape becomes inaccessible on the timescale of a simulation. For low enough temperatures, the configuration space becomes kinetically constrained to a subset of local minima, for which rearrangements are still feasible on the timescale in question. Glassy landscapes such as this will clearly make structure prediction difficult, since the relevant crystalline configuration space is well separated from the amorphous minima that are the likely result of an annealing simulation. Nevertheless, basin-hopping has tackled at least one glassy system successfully [134], and the prospects for future crystal structure prediction are intriguing.

References

1. Wales, D. J. (2003) *Energy Landscapes.* Cambridge University Press, Cambridge.
2. Born, M. and Oppenheimer, J. R. (1927) Zur quantentheorie der molekeln. *Ann. Physik*, **84**, 457–484.
3. Robb, M. A., Bernardi, F., and Olivucci, M. (1995) Conical intersections as a mechanistic feature of organic photochemistry. *Pure Appl. Chem*, **67**, 783–789.
4. Yarkony, D. R. (1998) Conical intersections: diabolical and often misunderstood. *Acc. Chem. Res.*, **31**, 511.
5. Allen, M. P. and Tildesley, D. J. (1987) *Computer Simulation of Liquids.* Clarendon Press, Oxford.
6. Frenkel, D. and Smit, B. (2002) *Understanding Molecular Simulation*, second edition. Academic Press, London.
7. Murrell, J. N. and Laidler, K. J. (1968) Symmetries of activated complexes. *Trans. Faraday. Soc.*, **64**, 371.
8. Wales, D. J. and Doye, J. P. K. (1997) Global optimization by basin-hopping and the lowest energy structures of Lennard–Jones clusters containing up to 110 atoms. *J. Phys. Chem. A*, **101**, 5111.
9. Doye, J. P. K. and Wales, D. J. (1998) Thermodynamics of global optimization. *Phys. Rev. Lett.*, **80**, 1357–1360.
10. Wales, D. J. and Scheraga, H. A. (1999) Global optimization of clusters, crystals and biomolecules. *Science*, **285**, 1368–1372.
11. McGinty, D. J. (1971) Vapor phase homogenous nucleation and the thermodynamic properties of small clusters of argon atoms. *J. Chem. Phys.*, **55**, 580.
12. Burton, J. J. (1972) Vibrational frequencies and entropies of small clusters of atoms. *J. Chem. Phys.*, **56**, 3133.
13. Hoare, M. R. (1979) Structure and dynamics of simple microclusters. *Adv. Chem. Phys.*, **40**, 49.
14. Stillinger, F. H. and Weber, T. A. (1984) Packing structures and transitions in liquids and solids. *Science*, **225**, 983.
15. Franke, G., Hilf, E. R., and Borrmann, P. (1993) The structure of small clusters: multiple normal-modes model. *J. Chem. Phys.*, **98**, 3496.
16. Wales, D. J. (1993) Coexistence in small inert-gas clusters. *Mol. Phys.*, **78**, 151.
17. Wales, D. J., Doye, J. P. K, Miller, M. A., Mortenson, P. N., and Walsh, T. R. (2000) Energy landscapes: from clusters to biomolecules. *Adv. Chem. Phys.*, **115**, 1–111.
18. Strodel, B. and Wales, D. J. (2008) Free energy surfaces from an extended harmonic superposition approach and kinetics for alanine dipeptide. *Chem. Phys. Lett.*, **466**, 105–115.
19. Krivov, S. V. and Karplus, M. (2002) Free energy disconnectivity graphs: application to peptide models. *J. Chem. Phys.*, **117**, 10894–10903.
20. Wales, D. J. (2002) Discrete path sampling. *Mol. Phys.*, **100**, 3285–3305.
21. Chekmarev, S. F., Krivov, S. V., and Karplus, M. (2005) Folding time distributions as an approach to protein folding kinetics. *J. Phys. Chem. B*, **109**, 5312–5330.
22. Noé, F., Krachtus, D., Smith, J. C., and Fischer, S. (2006) Transition networks for the comprehensive characterisation of complex conformational change in proteins. *J. Chem. Theory Comput.*, **2**, 840–857.
23. Noé, F., Oswald, M., Reinelt, G., Fischer, S., and Smith, J. C. (2006) Computing best transition pathways in high-dimensional systems: application to the $\alpha_l \rightleftharpoons \beta \rightleftharpoons \alpha_r$ transitions in octaalanine. *Multiscale Model Sim.*, **5**, 393–419.
24. Wales, D. J. (2006) Energy landscapes: calculating pathways and rates. *Int. Rev. Phys. Chem.*, **25**, 237–282.
25. Boulougouis, G. C. and Theodorou, D. N. (2007) Dynamical integration of a Markovian web: a first passage time approach. *J. Chem. Phys.*, **127**, 084903.
26. Noé, F. and Fischer, S. (2008) Transition networks for modeling the kinetics

of conformational change in macromolecules. *Curr. Op. Struct. Biol.*, **18**, 154–162.

27. Trygubenko, S. A. and Wales, D. J. (2004) A doubly-nudged elastic band method for finding transition states. *J. Chem. Phys.*, **120**, 2082.

28. Henkelman, G. and Jónsson, H. (1999) A dimer method for finding saddle points on high dimensional potential surfaces using only first derivatives. *J. Chem. Phys.*, **111**, 7010–7022.

29. Henkelman, G., Uberuaga, B. P., and Jónsson, H. (2000) A climbing image nudged elastic band method for finding saddle points and minimum energy paths. *J. Chem. Phys.*, **113**, 9901–9904.

30. Henkelman, G. and Jónsson, H. (2000) Improved tangent estimate in the nudged elastic band method for finding minimum energy paths and saddle points. *J. Chem. Phys.*, **113**, 9978–9985.

31. Munro, L. J. and Wales, D. J. (1999) Defect migration in crystalline silicon. *Phys. Rev. B*, **59**, 3969–3980.

32. Kumeda, Y., Munro, L. J., and Wales, D. J. (2001) Transition states and rearrangement mechanisms from hybrid eigenvector-following and density functional theory. Application to $C_{10}H_{10}$ and defect migration in crystalline silicon. *Chem. Phys. Lett.*, **341**, 185–194.

33. Pelzer, H. and Wigner, E. (1932) *Z. Phys. Chem.*, **B15**, 445.

34. Forst, W. (1973) *Theory of Unimolecular Reactions*. Academic Press, New York.

35. Miller, W. H. (1976) Importance of nonseparability in quantum mechanical transition-state theory. *Acc. Chem. Res.*, **9**, 306–312.

36. Wales, D. J. (2004) Some further applications of discrete path sampling to cluster isomerization. *Mol. Phys.*, **102**, 891–908.

37. Carr, J. M. and Wales, D. J. (2005) Global optimization and folding pathways of selected alpha-helical proteins. *J. Chem. Phys.*, **123**, 234901.

38. Strodel, B., Whittleston, C. S., and Wales, D. J. (2007) Thermodynamics and kinetics of aggregation for the GNNQQNY peptide. *J. Amer. Chem. Soc.*, **129**, 16005–16014.

39. Carr, J. M. and Wales, D. J. (2008) Folding pathways and rates for the three-stranded beta-sheet peptide beta3s using discrete path sampling. *J. Phys. Chem. B*, **112**, 8760–8769.

40. Li, Z. and Scheraga, H. A. (1987) Monte Carlo-minimization approach to the multiple-minima problem in protein folding. *Proc. Natl. Acad. Sci. USA*, **84**, 6611–6615.

41. Wales, D. J. (2005) The energy landscape as a unifying theme in molecular science. *Phil. Trans. Roy. Soc. A*, **363**, 357–377.

42. Wales, D. J. and Bogdan, T. V. (2006) Potential energy and free energy landscapes. *J. Phys. Chem. B*, **110**, 20765–20776.

43. Sibani, P., Schön, J. C., Salamon, P., and Andersson, J. (1993) Emergent hierarchical structures in complex-system dynamics. *Europhys. Lett.*, **22**, 479.

44. Schön, J. C. (1996) Studying the energy hypersurface of multi-minima systems – the threshold and the lid algorithm. *Ber. Bunsenges. Phys. Chem.*, **100**, 1388.

45. Schön, J. C., Putz, H., and Jansen, M. (1996) Studying the energy hypersurface of continuous systems – the threshold algorithm. *J. Phys. Cond. Matt.*, **8**, 143.

46. Schön, J. C. (2002) Energy landscape of two-dimensional lattice polymers. *J. Phys. Chem. A*, **106**, 10886–10892.

47. Czerminski, R. and Elber, R. (1990) Reaction path study of conformational transitions in flexible systems: applications to peptides. *J. Chem. Phys.*, **92**, 5580–5601.

48. Becker, O. M. and Karplus, M. (1997) The topology of multidimensional potential energy surfaces: theory and application to peptide structure and kinetics. *J. Chem. Phys.*, **106**, 1495–1517.

49. Wales, D. J., Miller, M. A., and Walsh, T. R (1998) Archetypal energy landscapes. *Nature*, **394**, 758.

50. Miller, M. A. and Wales, D. J. (1999) Energy landscape of a model protein. *J. Chem. Phys.*, **111**, 6610–6616.

51. Altis, A., Otten, M., Nguyen, P. H., Hegger, R., and Stock, G. (2008) Construction of the free energy landscape of biomolecules via dihedral angle principal component analysis. *J. Chem. Phys.*, **128**, 245102.
52. Krivov, S. V. and Karplus, M. (2004) Hidden complexity of free energy surfaces for peptide (protein) folding. *Proc. Nat. Acad. Sci. USA*, **101**, 14766–14770.
53. Krivov, S. V. and Karplus, M. (2006) One-dimensional free-energy profiles of complex systems: progress variables that preserve the barriers. *J. Phys. Chem. B*, **110**, 12689–12698.
54. Muff, S. and Caflisch, A. (2008) Kinetic analysis of molecular dynamics simulations reveals changes in the denatured state and switch of folding pathways upon single-point mutation of a β-sheet miniprotein. *Proteins: Struct., Func. Bioinf.*, **70**, 1185–1195.
55. Komatsuzaki, T., Hoshino, K., Matsunaga, Y., Rylance, G. J., Johnston, R. L., and Wales, D. J. (2005) How many dimensions are required to approximate the potential energy landscape of a model protein? *J. Chem. Phys.*, **122**, 084714.
56. Evans, D. A. and Wales, D. J. (2003) Free energy landscapes of model peptides and proteins. *J. Chem. Phys.*, **118**, 3891–3897.
57. Shaffer, J. S. and Chakraborty, A. K. (1993) Dynamics of poly(methyl methacrylate) chains adsorbed on aluminum surfaces. *Macromolecules*, **26**, 1120–1136.
58. Doye, J. P. K. and Wales, D. J. (1995) Calculation of thermodynamic properties of small Lennard–Jones clusters incorporating anharmonicity. *J. Chem. Phys.*, **102**, 9659–9672.
59. Calvo, F., Doye, J. P. K., and Wales, D. J. (2001) Characterization of anharmonicities on complex potential energy surfaces: perturbation theory and simulation. *J. Chem. Phys.*, **115**, 9627–9636.
60. Calvo, F., Doye, J. P. K., and Wales, D. J. (2001) Quantum partition functions from classical distributions. Application to rare gas clusters. *J. Chem. Phys.*, **114**, 7312–7329.
61. Jones, J. E. and Ingham, A. E. (1925) On the calculation of certain crystal potential constants, and on the cubic crystal of least potential energy. *Proc. R. Soc. A*, **107**, 636–653.
62. Doye, J. P. K., Miller, M. A., and Wales, D. J. (1999) Evolution of the potential energy surface with size for Lennard–Jones clusters. *J. Chem. Phys.*, **111**, 8417–8428.
63. Stillinger, F. H. and Weber, T. A. (1985) Computer simulation of local order in condensed phases of silicon. *Phys. Rev. B*, **31**, 5262–5271.
64. Middleton, T. F. and Wales, D. J. (2001) Energy landscapes of model glass formers. *Phys. Rev. B*, **64**, 024205.
65. Cornell, W. D., Cieplak, P., Bayly, C. I., Gould, I. R., Merz, K. M., Ferguson, D. M., Spellmeyer, D. C., Fox, T., Caldwell, J. W., and Kollman, P. A. (1995) A second generation force field for the simulation of proteins, nucleic acids and organic molecules. *J. Am. Chem. Soc.*, **117**, 5179–5197.
66. Mortenson, P. N., Evans, D. A., and Wales, D. J. (2002) Energy landscapes of model polyalanines. *J. Chem. Phys.*, **117**, 1363–1376.
67. Kumeda, Y. and Wales, D. J. (2003) Ab initio study of rearrangements between C_{60} fullerenes. *Chem. Phys. Lett.*, **374**, 125–131.
68. Jorgensen, W. L., Chandrasekhar, J., Madura, J. D., Impey, R. W., and Klein, M. L. (1983) Comparison of simple potential functions for simulating liquid water. *J. Chem. Phys.*, **79**, 926–935.
69. Chartrand, G. (1985) *Introductory Graph Theory*. Dover, New York.
70. Bryngelson, J. D. and Wolynes, P. G. (1987) Spin glasses and the statistical mechanics of protein folding. *Proc. Natl. Acad. Sci. USA*, **84**, 7524–6915.
71. Goldstein, R. A., Luthey-Schulten, Z. A., and Wolynes, P. G. (1992) Optimal protein-folding codes from spin-glass theory. *Proc. Natl. Acad. Sci. USA*, **89**(11): 4918–4922.

72. Mackay, A. L. (1962) A dense non-crystallographic packing of equal spheres. *Acta Cryst.*, **15**, 916–918.
73. Walsh, T. R. and Wales, D. J. (1998) Relaxation dynamics of C_{60}. *J. Chem. Phys.*, **109**, 6691–6700.
74. Leopold, P. E., Montal, M., and Onuchic, J. N. (1992) Protein folding funnels: a kinetic approach to the sequence–structure relationship. *Proc. Natl. Acad. Sci. USA*, **89**, 8721–8725.
75. Bryngelson, J. D., Onuchic, J. N., Socci, N. D., and Wolynes, P. G. (1995) Funnels, pathways, and the energy landscape of protein folding: a synthesis. *Proteins: Struct., Func. Gen.*, **21**, 167–195.
76. Miller, M. A., Doye, J. P. K., and Wales, D. J. (1999) Structural relaxation in Morse clusters: energy landscapes. *J. Chem. Phys.*, **110**, 328–334.
77. Doye, J. P. K., Miller, M. A., and Wales, D. J. (1999) The double-funnel energy landscape of the 38-atom Lennard–Jones cluster. *J. Chem. Phys.*, **110**, 6896–6906.
78. Neirotti, J. P., Calvo, F., Freeman, D. L., and Doll, J. D. (2000) Phase changes in 38-atom Lennard–Jones clusters. I. a parallel tempering study in the canonical ensemble. *J. Chem. Phys.*, **112**, 10340.
79. Calvo, F., Neirotti, J. P., Freeman, D. L., and Doll, J. D. (2000) Phase changes in 38-atom Lennard–Jones clusters. II. A parallel tempering study of equilibrium and dynamic properties in the molecular dynamics and microcanonical ensembles. *J. Chem. Phys.*, **112**, 10350.
80. Steinhardt, P. J., Nelson, D. R., and Ronchetti, M. (1983) Bond-orientational order in liquids and glasses. *Phys. Rev. B*, **28**, 784–805.
81. van Duijneveldt, J. S. and Frenkel, D. (1992) Computer simulation study of free energy barriers in crystal nucleation. *J. Chem. Phys.*, **96**, 4655–4668.
82. Doye, J. P. K. and Wales, D. J. (1999) The dynamics of structural transitions in sodium chloride clusters. *J. Chem. Phys.*, **111**, 11070–11079.
83. Doye, J. P. K. (2000) Effect of compression on the global optimization of atomic clusters. *Phys. Rev. E*, **62**, 8753–8761.
84. White, R. P. and Mayne, H. R. (1998) An investigation of two approaches to basin hopping minimization for atomic and molecular clusters. *Chem. Phys. Lett.*, **289**, 463–468.
85. Xue, G. L. (1991) Molecular conformation on the CM-5 by parallel two-level simulated annealing. *J. Global Opt.*, **4**, 187–208.
86. Barrón, C., Gómez, S., and Romero, D. (1996) Archimedean polyhedron structure yields a lower energy atomic cluster. *Appl. Math. Lett.*, **9**, 75–78.
87. Deaven, D. M., Tit, N., Morris, J. R., and Ho, K. M. (1996) Structural optimization of Lennard–Jones clusters by a genetic algorithm. *Chem. Phys. Lett.*, **256**, 195–200.
88. Barrón, C., Gómez, S., and Romero, D. (1997) Lower energy icosahedral atomic clusters with incomplete core. *Appl. Math. Lett.*, **10**, 25–28.
89. Holland, J. (1975) *Adaptation in Natural and Artificial Systems*. University of Michigan Press, Ann Arbor.
90. Goldberg, D. E. (1989) *Genetic Algorithms in Search, Optimization, and Machine Learning*. Addison-Wesley, Reading, MA.
91. Hartke, B. (2001) Global geometry optimization of atomic and molecular clusters by genetic algorithms. In Spector, E. Goodman, A. Wu, W. B. Langdon, H.-M. Voigt, M. Gen, S. Sen, M. Dorigo, S. Pezeshk, M. Garzon, and E. Burke, editors, *Proceedings of the Genetic and Evolutionary Computation Conference, GECCO-2001*, page 1284. Morgan Kaufmann, San Francisco.
92. Hartke, B. L. (2002) Efficient global geometry optimization of atomic and molecular clusters. In Pintér, editor, *Nonconvex Optimization and its Applications*. Kluwer, Dordrecht.
93. Johnston, R. L. and Roberts, C. (2003) Genetic algorithms for the geometry optimization of clusters and nanoparticles. In H. Cartwright, and L. Sztandera, editors, *Soft Computing Approaches in Chemistry*, pp. 25–61. Physica-Verlag, Heidelberg.

94. Deaven, D. M. and Ho, K. M. (1995) Molecular geometry optimization with a genetic algorithm. *Phys. Rev. Lett.*, **75**, 288–291.
95. Turner, G. W., Tedesco, E., Harris, K. D. M., Johnston, R. L., and Kariuki, B. M. (2000) Implementation of Lamarckian concepts in a genetic algorithm for structure solution from powder diffraction data. *Chem. Phys. Lett.*, **321**, 183–190.
96. Liu, D. and Nocedal, J. (1989) On the limited memory BFGS method for large scale optimization. *Math. Prog.*, **45**, 503–528.
97. Wales, D. J. GMIN: A program for basin-hopping global optimisation. http://www-wales.ch.cam.ac.uk/software.html.
98. Wales, D. J. OPTIM: A program for geometry optimisation and pathway calculations. http://www-wales.ch.cam.ac.uk/software.html.
99. Wales, D. J. (1992) Basins of attraction for stationary-points on a potential-energy surface. *J. Chem. Soc. Faraday Trans.*, **88**, 653–657.
100. Wales, D. J. (1993) Locating stationary-points for clusters in Cartesian coordinates. *J. Chem. Soc. Faraday Trans.*, **89**, 1305–1313.
101. Hoffmann, K. H., Franz, A., and Salamon, P. (2002) Structure of best possible strategies for finding ground states. *Phys. Rev. E*, **66**, 046706.
102. Hansmann, U. H. E. and Wille, L. T. (2002) Global optimization by energy landscape paving. *Phys. Rev. Lett.*, **88**, 068105.
103. Goedecker, S. (2004) Minima hopping: an efficient search method for the global minimum of the potential energy surface of complex molecular systems. *J. Chem. Phys.*, **120**, 9911–9917.
104. Xiang, Y., Jiang, H., Cai, W., and Shao, X. (2004) An efficient method based on lattice construction and the genetic algorithm for optimization of large Lennard–Jones clusters. *J. Phys. Chem. A*, **108**, 3586–3592.
105. Shao, X., Xiang, Y., and Cai, W. (2005) Structural transition from icosahedra to decahedra of large Lennard–Jones clusters. *J. Phys. Chem. A*, **109**, 5193–5197.
106. Zhan, L., Chen, J. Z. Y., Liu, W.-K., and Lai, S. K. (2005) Asynchronous multicanonical basin-hopping method and its application to cobalt nanoclusters. *J. Chem. Phys.*, **122**, 244707.
107. Verma, A., Schug, A., Lee, K. H., and Wenzel, W. (2006) Basin-hopping simulations for all-atom protein folding. *J. Chem. Phys.*, **124**, 044515.
108. Zhan, L., Chen, J. Z. Y., and Liu, W. K. (2006) Monte Carlo basin paving: an improved global optimization method. *Phys. Rev. E*, **73**, 015701R.
109. Cheng, L. and Yang, J. (2007) Novel lattice-searching method for modeling the optimal strain-free close-packed isomers of clusters. *J. Phys. Chem. A*, **111**, 2336–2342.
110. Cheng, L. and Yang, J. (2007) Global minimum structures of Morse clusters as a function of the range of the potential: $81 \leq n \leq 160$. *J. Phys. Chem. A*, **111**, 5287–5293.
111. Kim, H. G., Choi, S. K., and Lee, H. M. (2008) New algorithm in the basin hopping Monte Carlo to find the global minimum structure of unary and binary metallic nanoclusters. *J. Chem. Phys.*, **128**, 144702.
112. Cheng, L., Feng, Y., Yang, J., and Yang, J. (2009) Funnel hopping: searching the cluster potential energy surface over the funnels. *J. Chem. Phys.*, **130**, 214112.
113. Schonborn, S. E., Goedecker, S., Roy, S., and Oganov, A. R. (2009) The performance of minima hopping and evolutionary algorithms for cluster structure prediction. *J. Chem. Phys.*, **130**, 144108.
114. Gehrke, R. and Reuter, K. (2009) Assessing the efficiency of first-principles basin-hopping sampling. *Phys. Rev. B*, **79**, 085412.
115. Rossi, G. and Ferrando, R. (2009) Searching for low-energy structures of nanoparticles: a comparison of different methods and algorithms. *J. Phys. Cond. Matt.*, **21**, 084208.
116. Froltsov, V. A. and Reuter, K. (2009) Robustness of 'cut and splice' genetic

algorithms in the structural optimization of atomic clusters. *Chem. Phys. Lett.*, **473**, 363–366.
117. Cvijovic, D. and Klinowski, J. (1995) Taboo search: an approach to the multiple minima problem. *Science*, **267**, 664–666.
118. Ji, M. and Klinowski, J. (2006) Taboo evolutionary programming: a new method of global optimisation. *Proc. Roy. Soc. A*, **462**, 3613–3627.
119. Zabrodsky, H., Peleg, S., and Avnir, D. (1992) Continuous symmetry measures. *J. Am. Chem. Soc.*, **114**, 7843–7851.
120. Katzenelson, O., Hel-Or, H. Z., and Avnir, D. (1996) chirality of large random supramolecular structures. *Chem. Eur. J.*, **2**, 174–181.
121. Wales, D. J. (1998) Symmetry, near-symmetry and energetics. *Chem. Phys. Lett.*, **285**, 330–336.
122. Wales, D. J. (1998) Erratum: Symmetry, near-symmetry and energetics (vol. 285, p 330). *Chem. Phys. Lett.*, **294**, 262.
123. Wales, D. J. and Hodges, M. P. (1998) Global minima of water clusters $(H_2O)_N$, $N \leq 21$, described by an empirical potential. *Chem. Phys. Lett.*, **286**, 65–72.
124. Hartke, B. (2003) Size-dependent transition from all-surface to interior-molecule structures in pure neutral water clusters. *Phys. Chem. Chem. Phys.*, **5**, 275–284.
125. Hartke, B. (2008) Morphing Lennard–Jones clusters to TIP4P water clusters: why do water clusters look like they do? *Chem. Phys.*, **346**, 286–294.
126. Day, G. M., Chisholm, J., Shan, N., Motherwell, S., and Jones, W. (2004) An assessment of lattice energy minimization for the prediction of crystal structures. *Cryst. Growth Des.*, **4**, 1327–1340.
127. Hodges, M. P. and Wales, D. J. (2000) Global minima of protonated water clusters. *Chem. Phys. Lett.*, **324**, 279–288.
128. James, T. and Wales, D. J. (2005) Protonated water clusters described by an empirical valence bond potential. *J. Chem. Phys.*, **122**, 134306.
129. James, T., Wales, D. J., and Hernández-Rojas, J. (2005) Global minima for water clusters $(H_2O)_N$, $N < 21$, described by a five-site empirical potential. *Chem. Phys. Lett.*, **415**, 302–307.
130. Kabrede, H. and Hentschke, R. (2003) Global minima of water clusters $(H_2O)_N$, $N \leq 25$, described by three empirical potentials. *J. Phys. Chem. B*, **107**, 3914–3920.
131. Kazachenko, S. and Thakkar, A. J. (2009) Improved minima-hopping. TIP4P water clusters, $(H_2O)_N$ with $N \leq 37$. *Chem. Phys. Lett.*, **476**, 120–124.
132. Doye, J. P. K. and Wales, D. J. (1999) Structural transitions and global minima of sodium chloride clusters. *Phys. Rev. B*, **59**, 2292–2300.
133. Leary, R. H. and Doye, J. P. K. (1999) Tetrahedral global minimum for the 98-atom Lennard–Jones cluster. *Phys. Rev. E*, **60**, R6320–R6322.
134. Middleton, T. F., Hernández-Rojas, J., Mortenson, P. N., and Wales, D. J. (2001) Crystals of binary Lennard–Jones solids. *Phys. Rev. B*, **64**, 184201.
135. Rapacioli, M., Calvo, F., Spiegelman, F., Joblin, C., and Wales, D. J. (2005) Stacked clusters of polycyclic aromatic hydrocarbon molecules. *J. Phys. Chem. A*, **109**, 2487–2497.
136. Miller, M. A. and Wales, D. J. (2005) Novel structural motifs in clusters of dipolar spheres: knots, links, and coils. *J. Phys. Chem. B*, **109**, 23109–23112.
137. Wales, D. J. and Ulker, S. (2006) Structure and dynamics of spherical crystals characterized for the Thomson problem. *Phys. Rev. B*, **74**, 212101.
138. Wales, D. J., McKay, H., and Altschuler, E. L. (2009) Defect motifs for spherical topologies. *Phys. Rev. B*, **79**, 224115.
139. Wales, D. J., Doye, J. P. K., Dullweber, A., Hodges, M. P., Naumkin, F. Y., Calvo, F., Hernández-Rojas, J., and Middleton, T. F. The Cambridge cluster database, http://www-wales.ch.cam.ac.uk/ccd.html.
140. Doye, J. P. K. and Wales, D. J. (1997) Structural consequences of the range

of the interatomic potential – a menagerie of clusters. *J. Chem. Soc., Faraday Trans.*, **93**, 4233–4243.
141. Doye, J. P. K. and Wales, D. J. (1998) Global minima for transition metal clusters described by Sutton–Chen potentials. *New J. Chem.*, **22**, 733–744.
142. Hernández-Rojas, J., Breton, J., Llorente, J. M. G., and Wales, D. J. (2006) Global potential energy minima of $C_{60}(H_2O)_N$ clusters. *J. Phys. Chem. B*, **110**, 13357–13362.
143. Naumkin, F. Y. and Wales, D. J. (2002) Diatomics-in-molecules potentials incorporating *ab initio* data: application to ionic, Rydberg-excited, and molecule-doped rare gas clusters. *Comput. Phys. Comm.*, **145**, 141–155.
144. Thomson, J.J. (1904) On the structure of the atom. *Philos. Mag.*, **7**, 237–265.
145. Bausch, A. R., Cacciuto, A. Bowick, M. J., Dinsmore, A. D., Hsu, M. F., Nelson, D. R., Nikolaides, M. G., Travesset, A., and Weitz, D. A. (2003) Grain boundary scars and spherical crystallography. *Science*, **299**, 1716–1718.
146. Pérez-Garrido, A. and Moore, M. A. (1999) Symmetric patterns of dislocations in Thomson's problem. *Phys. Rev. B*, **60**, 15628–15631.
147. Bowick, M. J., Nelson, D. R., and Travesset, A. (2000) Interacting topological defects on frozen topologies. *Phys. Rev. B*, **62**, 8738–8751.
148. Kroto, H. W., Heath, J. R., O'Brien, S. C., Curl, R. F., and Smalley, R. E. (1985) C_{60}: buckminsterfullerene. *Nature*, **318**, 162–163.
149. Pérez-Garrido, A. (2000) Giant multilayer fullerene structures with symmetrically arranged defects. *Phys. Rev. B*, **62**, 6979–6981.
150. Caspar, D. L. D. and Klug, A. (1962) Physical principles in the construction of regular viruses. *Cold Spring Harbour, Symp. Quant. Biol.*, **27**, 1–24.
151. Marzec, C. J. and Day, L. A. (1993) Pattern formation in icosahedral virus capsids: the papova viruses and nudaurelia capensis beta virus. *Biophys. J.*, **65**, 2559–2577.
152. Cho, Y.-S., Yi, G.-R., Lim, J.-M., Kim, S.-H., Manoharan, V. N., Pine, D. J., and Yang, S.-M. (2005) Self-organization of bidisperse colloids in water droplets. *J. Am. Chem. Soc.*, **127**, 15968–15975.
153. Dodgson, M. J. W. (1996) Investigation on the ground states of a model thin film superconductor on a sphere. *J. Phys. A*, **29**, 2499–2508.
154. Dodgson, M. J. W. and Moore, M. A. (1997) Vortices in a thin-film superconductor with a spherical geometry. *Phys. Rev. B*, **55**, 3816–3830.
155. Simons, K. and Vaz, W. L. C. (2004) Model systems, lipid rafts, and cell membranes. *Annu. Rev. Biophys. Biomol. Struct.*, **33**, 269–295.
156. Masuda, Y., Itoh, T., and Koumoto, K. (2005) Self-assembly and micropatterning of spherical-particle assemblies. *Adv. Mater.*, **17**, 841–845.
157. Lipowsky, P., Bowick, M. J., Meinke, J. H., Nelson, D. R., and Bausch, A. R. (2005) Direct visualization of dislocation dynamics in grain-boundary scars. *Nat. Mater.*, **4**, 407–411.
158. Einert, T., Lipowsky, P., Schilling, J., Bowick, M. J., and Bausch, A. R. (2005) Grain boundary scars on spherical crystals. *Langmuir*, **21**, 12076–12079.
159. Pum, D., Messner, P., and Sleytr, U. B. (1991) Role of the s layer in morphogenesis and cell division of the archaebacterium methanocorpusculum sinense. *J. Bacteriol.*, **173**, 6865–6873.
160. Sleytr, U. B., Sára, M., Pum, D., and Schuster, B. (2001) Characterization and use of crystalline bacterial cell. Surface layers. *Prog. Surf. Sci.*, **68**, 231–278.
161. Albrecht, U. and Leiderer, P. (1992) On the correlation between properties of multielectron dimples and bubbles. *J. Low Temp. Phys.*, **86**, 131–251.
162. Leiderer, P. (1995) Ions at helium interfaces. *Z. Phys. B*, **98**, 303–308.
163. Day, G. M., Motherwell, S., and Jones, W. (2005) Beyond the isotropic atom model in crystal structure prediction of rigid molecules: atomic multipoles versus point charges. *Cryst. Growth Des.*, **5**, 1023–1033.
164. Oganov, A. R. and Glass, C. W. (2006) Crystal structure prediction using ab

initio evolutionary techniques: principles and applications. *J. Chem. Phys.*, **124**, 244704.

165. Abraham, N. L. and Probert, M. I. J. (2006) A periodic genetic algorithm with real-space representation for crystal structure and polymorph prediction. *Phys. Rev. B*, **73**, 224104.

166. Day, G. M., Motherwell, S., and Jones, W. (2007) A strategy for predicting the crystal structures of flexible molecules: the polymorphism of phenobarbital. *Phys. Chem. Chem. Phys.*, **9**, 1693–1704.

167. Woodley, S. M. and Catlow, R. (2007) Crystal structure prediction from first principles. *Nature Mater.*, **7**, 937–946.

168. Price, S. (2008) From crystal structure prediction to polymorph prediction: interpreting the crystal energy landscape. *Phys. Chem. Chem. Phys.*, **10**, 1996–2009.

169. Price, S. L. (2009) Computed crystal energy landscapes for understanding and predicting organic crystal structures and polymorphism. *Accounts Chem. Res.*, **42**, 117–126.

170. Doye, J. P. K., Wales, D. J., and Miller, M. A. (1998) Thermodynamics and the global optimization of Lennard–Jones clusters. *J. Chem. Phys.*, **109**, 8143–8153.

171. Calvo, F., Bogdan, T. V., de Souza, V. K., and Wales, D. J. (2007) Equilibrium density of states and thermodynamic properties of a model glass former. *J. Chem. Phys.*, **127**, 044508.

172. de Souza, V. K. and Wales, D. J. (2008) Energy landscapes for diffusion: analysis of cage-breaking processes. *J. Chem. Phys.*, **129**, 164507.

173. Kob, W. and Andersen, H. C. (1994) Scaling behavior in the beta-relaxation regime of a supercooled Lennard–Jones mixture. *Phys. Rev. Lett.*, **73**, 1376–1379.

174. Kob, W. and Andersen, H. C. (1995) Testing mode-coupling theory for a supercooled binary Lennard–Jones mixture: the Van Hove correlation function. *Phys. Rev. E*, **51**, 4626–4641.

175. Kob, W. and Andersen, H. C. (1995) Testing mode-coupling theory for a supercooled binary Lennard–Jones mixture. II. intermediate scattering function and dynamic susceptibility. *Phys. Rev. E*, **52**, 4134–4153.

176. Sastry, S., Debenedetti, P. G., and Stillinger, F. H. (1998) Signatures of distinct dynamical regimes in the energy landscape of a glass-forming liquid. *Nature*, **393**, 554–557.

177. Büchner, S. and Heuer, A. (1999) Potential energy landscape of a model glass former: thermodynamics, anharmonicities, and finite size effects. *Phys. Rev. E*, **60**, 6507–6518.

178. Sciortino, F., Kob, W., and Tartaglia, P. (1999) Inherent structure entropy of supercooled liquids. *Phys. Rev. Lett.*, **83**, 3214–3217.

179. Middleton, T. F. and Wales, D. J. (2003) Energy landscapes of model glass formers. II. Results for constant pressure. *J. Chem. Phys.*, **118**, 4583–4593.

180. Sciortino, F. and Tartaglia, P. (2001) Extension of the fluctuation-dissipation theorem to the physical aging of a model glass-forming liquid. *Phys. Rev. Lett.*, **86**, 107–110.

181. Sastry, S. (2001) The relationship between fragility, configurational entropy and the potential energy landscape of glass-forming liquids. *Nature*, **409**, 164–167.

182. Stoddard, S. D. and Ford, J. (1973) Numerical experiments on the stochastic behavior of a Lennard–Jones system. *Phys. Rev. A*, **8**, 1504–1512.

183. Weber, T. A. and Stillinger, F. H. (1985) Local order and structural transitions in amorphous metal-metalloid alloys. *Phys. Rev. B*, **31**, 1954–1963.

184. Fernández, J. R. and Harrowell, P. (2003) Crystal phases of a glass-forming Lennard–Jones mixture. *Phys. Rev. E*, **67**, 011403.

185. Swendsen, R. H. and Wang, J.-S. (1986) Replica Monte Carlo simulation of spin-glasses. *Phys. Rev. Lett.*, **57**, 2607–2609.

186. Geyer, G. (1991) Markov chain Monte Carlo maximum likelihood. In Keramidas, editor, *Computing Science and Statistics: Proceedings of the 23rd Symposium on the Interface*,

187. Ferrenberg, A. M. and Swendsen, R. H. (1988) New Monte Carlo technique for studying phase transitions. *Phys. Rev. Lett.*, **61**, 2635–2638.
188. Ferrenberg, A. M. and Swendsen, R. H. (1989) Optimized Monte Carlo data analysis. *Phys. Rev. Lett.*, **63**, 1195–1198.
189. Calvo, F. and Labastie, P. (1995) Configurational density of states from molecular dynamics simulations. *Chem. Phys. Lett.*, **247**, 395–400.
190. Anderson, P. W. (1979) In Balian, R. Maynard, and G. Toulouse, editors, *Ill-Condensed Matter*, pp. 159–261. North-Holland, Amsterdam.
191. Dyre, J. C. (2006) The glass transition and elastic models of glass-forming liquids. *Rev. Mod. Phys.*, **78**, 953–967. pp. 156–163. Inferface Foundation, Fairfax Station.
192. de Souza, V. K. and Wales, D. J. (2009) Connectivity in the potential energy landscape for binary Lennard–Jones systems. *J. Chem. Phys.*, **130**, 194508.
193. de Souza, V. K. and Wales, D. J. (2005) Diagnosing broken ergodicity using an energy fluctuation metric. *J. Chem. Phys.*, **123**, 134504.
194. Mountain, R. D. and Thirumalai, D. (1989) Measures of effective ergodic convergence in liquids. *J. Phys. Chem.*, **93**, 6975–6979.
195. Stillinger, F. H. (1995) A topographic view of supercooled liquids and glass formation. *Science*, **267**, 1935–1939.
196. Doliwa, B. and Heuer, A. (2003) Hopping in a supercooled Lennard–Jones liquid: metabasins, waiting time distribution, and diffusion. *Phys. Rev. E*, **67**, 030501R.

3
Random Search Methods
William W. Tipton and Richard G. Hennig

3.1
Introduction

A primary goal of computational and theoretical materials engineering is the identification of materials with desirable properties. Often, we have an application in mind and can describe the properties of a material which may be successfully applied to our problem, e.g., it is light weight, cheap, strong, insulating, or has specific bandgap or diffusion coefficients. It is then up to a materials engineer to find such a material.

To this end, we often approach the inverse problem. That is, instead of starting with a list of properties and working directly to a material solution, we start by considering a particular material and try to determine its properties. This is easier. Of course, once the properties of a long list of materials are known, it is likely we will be able to select from the list materials which satisfy the constraints of a given application [1, 2].

It turns out that parts of this problem are, today, very routine calculations. *Once we know the atomistic structure of a material,* methods such as density functional theory (DFT) implemented in a number of mature software packages allow us to predict a material's electronic structure, elastic constants, etc. [3].

However, the question of how to find a material's atomistic structure is an open one, and thus the need for the present text. Indeed, the question itself needs to be more precisely specified since a material's structure may depend probabilistically on growth conditions, its processing history, etc. We know that at thermodynamic equilibrium, a material will take on the structure with the lowest free energy given by

$$G = U - TS + PV$$

where U, T, S, P, and V are the internal energy, temperature, entropy, pressure, and volume of the system, respectively.

Since we compare trial solutions to each other in order to find the lowest energy structure, we are not as interested in the absolute free energy of any particular structure but in *differences* of free energies. The PV term is easy to find and include in a calculation and has significant effect on the results primarily when one is

Modern Methods of Crystal Structure Prediction. Edited by Artem R. Oganov
Copyright © 2011 WILEY-VCH Verlag GmbH & Co. KGaA, Weinheim
ISBN: 978-3-527-40939-6

studying systems under high pressure. In practice, the internal energy U will account for most of the energy difference between phases (as well as most of the algorithm's run time). It may be calculated by way of energy models such as empirical potentials (using, e.g., GULP [4]) or DFT (using, e.g., PWSCF [5]).

We can break the entropy term S into three contributions: electronic, configurational, and vibrational entropy [6].

$$S = S_{el} + S_{conf} + S_{vib}$$

The electronic term S_{el} is relatively easy to calculate but typically of negligible magnitude. The configurational and vibrational contributions can be significant but are difficult to compute as they require extensive sampling of the potential energy surface. For these reasons, the entropic terms in the free energy are often neglected in these calculations. This is often safe to do since we are primarily interested in differences in energies rather than the absolute quantities, and the S contributions cancel to some extent between different phases. However, there is the possibility of entropic stabilization in which structures which are not even mechanically stable at zero temperature can be stabilized entropically [7, 8].

Nonetheless, due to the computational cost of directly estimating the free energy, stochastic-search algorithms are generally not applied to the high-temperature problem (Monte Carlo or Molecular Dynamics techniques may be useful here). We neglect the entropy for simplicity, essentially confining ourselves to the zero temperature regime. In this case, a material's free energy is simply its enthalpy $H = U + PV$, and a material's thermodynamically stable crystal structure is that arrangement of atoms which has the lowest enthalpy. Hence, to find the physically realized crystal structure, we must search for the one with the lowest enthalpy.

In this light, we are viewing atomic structure prediction as an optimization problem. That is, if we view the energy of a system as a function of various parameters describing it (atomic positions, etc.), then predicting the stable structure is equivalent to finding the values of the parameters that minimize the energy function.

Unfortunately, the energy functions of real systems are not simple objects and we cannot write analytic expressions for them. These functions themselves are expensive to compute, and high-order derivatives thereof are prohibitively so. Thus, our functions are not amenable to many traditional optimization methods to find the global minimum. Methods which rely on high-order derivatives may be immediately discounted. Deterministic search strategies such as branch-and-bound may be considered, but they remain exceedingly computationally expensive even with the simplest of energy models [9].

However, any optimization problem for which we can describe and evaluate solutions is amenable to one of the most simple optimization schemes: guess and check. We guess a variety of possible solutions, evaluate the quality of each of them, and choose the best. There are many ways one can imagine to guess solutions including researcher intuition, but one of the most simple choices is the topic of this chapter: random search. Additionally, we will see that local minimization routines improve random searches beyond the trivial guess and check methodology.

We begin the remainder of this chapter by briefly discussing the history of random search methods and their application to crystal structures. We present an overview of the theory motivating the design of most methods. Finally, we survey the method's application to various systems of interest in the literature.

3.2 History and Overview

There are many optimization problems of great practical importance, and random search algorithms have long been applied to their solution. The method was probably [10] first suggested by Anderson [11] in the context of operations research and further investigated shortly thereafter by Rastrigin [12] and Karnopp [13]. More recent texts by Spall [14] and Zhigljavsky [15] provide a comprehensive discussions of stochastic search and optimization methods. Before looking at the details of any particular implementation of a random search algorithm for atomic structure prediction, we make some general comments about the method.

Given some details about our system of interest, say the stoichiometry of a solid crystal or the sequence of a protein, the random search program is to repeatedly generate some random arrangement of the system's atomic or molecular components and, subject to local minimization, compute the energy of that arrangement. This is repeated until a sufficient solution is found. We can diagram this simple algorithm as follows:

1) randomly generate structure;
2) apply local optimization routine to minimize structure's energy;
3) repeat until convergence.

It is obvious that randomly guessing solutions is not the most efficient way to solve the problem since the method does not attempt to leverage anything we know about the system a priori or anything we learn about it over the course of the method. However, the simplicity of this approach which is the root of its weaknesses also leads to several advantages.

Firstly, the simplicity of the algorithm means that it is relatively quick and easy to implement. As computer time becomes less and less expensive relative to programmer time, navigating this trade-off becomes more important.

Secondly, the method need only use the energy of particular structures. Random search methods commonly use low-order derivatives, i.e., forces and stresses, for local optimization, but they are not required, and so the method is compatible with energy routines which either cannot produce such extra information or are prohibitively slow in doing so.

Next, although determining convergence to the absolute ground state is challenging, we find in practice that the method is often very quick to find good solutions, energetically low-lying configurations which bear many similarities to the thermodynamic minimum and which may occur as metastable phases in the material. As we will see in specific cases below, with the significant reduction in problem complexity afforded us by local optimization, we can also often be confident that the algorithm has found the global minimum.

Finally, the method is easily amenable to statistical analysis for describing its convergence properties. Anderssen and Bloomfield show that random searching is more effective than searching on a uniform grid in phase spaces of dimensionality greater than six [16]. Spall shows how to estimate the number of trials required to assure a certain probability of sampling the correct solution. He assumes that a finite volume of the solution phase space corresponds to the optimal solution (such as the basin of attraction of the local minimizer in our problem) and derives an expression for the number of samples required to guarantee a certain probability of finding it [14]. See [10] for discussion of the method's convergence rate and [14] for additional mathematical details. More on the method's convergence rate and mathematical details can be found in Refs. [10] and [14], respectively.

The most computationally expensive step in this process is the energy calculation. Although the details of the system, algorithm, and energy method are relevant to the method's success, accurate energy calculations typcially take several orders of magnitude more computer time than any other step in the algorithm. For this reason, run-time of the algorithm may be described by the number of total energy calculations which must be performed to achieve convergence, and it is important to minimize this number.

3.3
Methods

It is necessary to specify the form of trial solutions. This choice has wide consequences for the success of the algorithm. Most work on structure prediction by random search has focused on crystalline solids. We will concentrate on that problem. Similar considerations will apply when parameterizing other types of systems such as molecules or nanoclusters.

By assuming periodic boundary conditions, an infinite crystal may be specified by a Bravais lattice and a basis of atoms. The lattice may be specified by six lattice parameters (such as three angles and three lengths), and the basis by N atomic coordinates (i.e., $3N - 3$ numbers for 3D crystals, taking into account translational invariance). Thus, the solution phase space for this problem is of $3N + 3$ dimensions. The number of atoms in the basis, N, is a parameter which itself may need to be determined by the search algorithm.

The solution space we have described is high dimensional and infinite in extent in most of the dimensions. However, many points in the phase space either represent clearly unphysical structure or are redundant, describing crystals which are also described by other points in the space or represent crystals which may be immediately excluded as unphysical. By designing some simple criteria that eliminate the obviously nonphysical structures and confining our search to a single representation of each crystal, one can avoid considering large portions of the total solution space and simplify the search problem.

Unphysical solutions include those which contain atoms spread very far apart. By applying constraints on the parameters of our trial solutions: minimum and maximum lattice parameters, we confine our search to a bounded space. We may also constrain the range of allowable nearest-neighbor distances or crystal densities to further narrow our search to physically realizable structures. Incidentally, this often also helps to ensure the stability of energy codes.

Now, notice that even the bounded solution space is highly redundant. A single physical crystal can be represented by many different unit cell choices, by shifting all the atoms by some constant amount, or by swapping the coordinates of two identical atoms in the basis. For clusters or molecules, rotations are also redundant degrees of freedom. It is important that search algorithms attempt to represent each structure in a single, standardized way. In this way, the algorithm may avoid redundant calculations and, again, significantly reduce the effective size of the solution space. We may also constrain trial solutions based on experimentally known data such as the space group or structural motifs (such as H_2O units in ice).

To this end the algorithm generally enforces that a particular species is located at the origin in crystal coordinates and that the lattice itself is chosen in a standardized form. One such form is described by Pauschenwein [17]. He presents the construction of "a general parameterization for all three-dimensional crystal lattices ... which guarantees that the three primitive vectors constructed by the parametrization are the three shortest possible, linearly independent lattice vectors existing in the whole lattice" [17]. This Minimum Distance Parameterization removes almost all of the redundancy in the phase space.

The most important technique for reducing the complexity of the search problem is local relaxation of trial structures. While the traditional optimization algorithms built into most energy codes cannot automatically find the *global* energy minimum, they can efficiently *relax* a given structure to a nearby *local* minimum. The particular methods used include conjugate gradient optimization, several quasi-Newton methods, and damped molecular dynamics [18, 19]. Essentially, the energy determines the forces acting on each atom and moves it "downhill" until the system reaches a minimum of the energy surface. Notice that this will not usually overcome any energy barriers between the trial solution and the true ground state. This is illustrated in Figure 3.1 using a one-dimensional function. Note that the potential energy surfaces of real systems lie in spaces of high dimensionality which leads to complications not apparent in the 1D case.

Performing the relaxation algorithm on each trial structure before calculating its energy significantly reduces the complexity of the structure search problem. Instead of randomly guessing the ground state solution itself, the method must only guess a solution "nearby" the ground state. We may think of this as partitioning the solution phase space into regions of attraction of the local optimization method surrounding each local minimum. This is illustrated in Figure 3.2. To completely search the solution space, we must no longer sample every point in the space, but merely one point in each region [20].

With this technique, we may begin to speak with some confidence of exhaustively sampling a space. A properly constrained solution space may have few enough

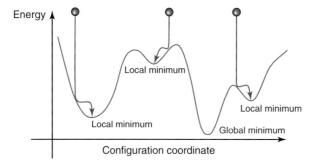

Figure 3.1 The random search process using local minimization. The algorithm randomly generates a variety of trial solutions which are relaxed to a nearby local minimum using a traditional optimization routine.

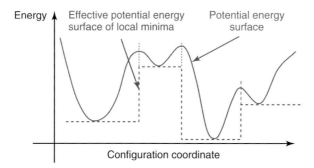

Figure 3.2 Effective potential energy surface. The use of a local minimizer simplifies the search problem by transforming the continuous solution space with infinitely many trial solutions into a discrete space with finitely many if we constrain our search space to a finite volume using physical constraints.

local minima that a search algorithm may sample all of them in a reasonable amount of time. Several authors suggest that their random searches may have indeed been nearly or fully exhaustive since they found many structures several times each [21, 22].

However these works considered relatively small unit cells. As per our discussion above, the size of the solution space and the complexity of our problem may be quantified by the number of local minima in the constrained space. The dimension of the search space grows linearly with the N. Therefore, it is believed that the number of local minima in the solution space and, thus, the number of trial solutions we need to adequately sample it, grows exponentially with N [23, 24].

Finally, a search method must specify criteria for algorithm convergence. By the nature of the method, it can never find the ground state solution with absolute certainty. This limitation is, of course, common to any other search method, such as genetic algorithms. In practice, several naive convergence criteria (or "stopping criteria") work well. A search is usually considered converged when no improvement in the best trial solution has been made over several iterations and the current best solution has been found several times [25].

Venkatesh et al. have developed a more sophisticated statistical method to this end [26]. The method applies Bayesian analysis to the set of local minima found over the course of the random search to approximate the distribution of the number of local minima. From this, they derive the convergence criterion which tries to navigate the computer time/solution confidence trade-off.

3.4
Applications and Results

Due to their periodic structure, simple crystalline solids generally have many fewer degrees of freedom than do, e.g., proteins. Thus, work making use of random structure searches has focused on these simple systems where the computer time/programmer time trade-off is appropriate. In all cases, the energy and local minimization routines used are standard, so a work's random structure generation method and convergence criteria are its salient features. Some of the earliest work on structure prediction by random search gave a proof-of-concept on Lennard–Jones systems [27]. The following works describe real materials.

The Pickard and Needs group from Cambridge has been very prolific in their use of the random search strategy in recent years having investigated a number of interesting molecular crystal and semiconductor systems described by DFT. In 2006 they described the first application of their method to silane (SiH_4) at high pressures in order to find phases which may superconduct [22]. They argue that the method is particularly suitable for systems under high pressure since such phases often have simple structures in the sense that they frequently have small unit cells. They perform their search separately for cells of different numbers of atoms, N. Once the number and type of atoms in the cell is chosen, they stochastically generate a lattice by selecting three lattice lengths distributed uniformly on [0.5, 1.5] and three lattice angles distributed uniformly on [40°, 140°]. The volume of the entire cell is then scaled to between 0.5 and 1.5 of some given volume, and Si and H atoms are given random coordinates uniformly distributed in the cell. They found a metallic phase which should be accessible experimentally.

In the same year, the group studied high-pressure structures of CaC_6, a material whose superconducting properties vary with pressure [25]. Initial structures for the search were specified to contain seven atoms and the lattice randomly generated such that the density was within a factor of two of the known low-pressure phase. The authors suggest that the search was likely nearly exhaustive (over structures with $N = 7$) since they found many phases several times. They used intuition gained from the seven-atom case to construct and test structures of larger N. The study yielded several structures which were favored at different pressures and which were later experimentally confirmed [28].

The same authors have modified the random search technique to find the structure of hydrogen defects in silicon. This was done by enforcing additional constraints on the trials solutions, i.e., each trial solution contained a defect. They found novel structures for defect clusters of various sizes [29]. Similar studies of high-pressure phases of solid hydrogen [30], nitrogen [31], lithium [32], H_2O [33], aluminum hydride [34], and iron [35] yielded novel results. A review by Pickard and Needs of their own work is given in [36] and provides additional insight into their approach.

Feng et al. performed an extensive structural search of Li–Be compounds using both random search and additional guess and check based on chemical intuition [21]. The random search was instrumental in identifying stable high-pressure phases where the researchers' intuition was less successful. They found four stoichiometric Li_xBe_{1-x} compounds stable over a range of pressures, several of whc display quite unusual and unexpected electronic properties. In Figure 3.3, we show the structures and the pressure ranges over which they are stable.

Successful random search work has been performed on metal and metal–alloy nanoclusters by Johnston et al. In Ref. [37] they study aluminum clusters described by the Murrell–Mottram potential. Trial clusters are randomly generated subject to constraints on minimum and maximum nearest-neighbor distance and the constraint that all atoms lie within a sphere of radius proportional to the number of atoms $N^{\frac{1}{3}}$. The clusters are then relaxed using a quasi-Newton method. In this way, 1000 trial solutions were prepared and tested for clusters of each size $N = 2, \ldots, 20$. In most cases, the ground state is found within the first 100 trials and on the order of 10 times in total. Additionally, they describe evidence for exhaustive sampling in some of these clusters. However, they also find that the method becomes less effective with increasing system size [37].

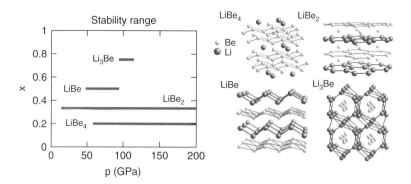

Figure 3.3 Li–Be compounds found at a high pressure using a random search method by Feng et al. [21]. (Please find a color version of this figure on the color plates.)

Table 3.1 Summary of random structure search applications.*

Study	System	Ground state	System size N	Energy model
Schon et al. 1994 [27]	Lennard–Jones	HCP/FCC	8 atoms	Lennard–Jones
Wang et al. 1999 [41]	Protein complexes	Various	Large	AMBER potential
Lloyd et al. 1998 [37]	Al clusters	Novel structures for various N	Up to 20	MM potential
Bailey et al. 2003 [38]	Ni–Al clusters	Novel structures for various N	Up to 55	Gupta potential
Pickard et al. 2007 [30]	Solid hydrogen	Novel high-pressure structures	Up to 24	DFT
Pickard et al. 2006 [22]	Silane	Various high-pressure phases	Up to 10	DFT
Pickard et al. 2009 [32]	Lithium	Various high-pressure phases	Up to 24	DFT
Pickard et al. 2009 [31]	Nitrogen	Various high-pressure phases	Up to 12	DFT
Pickard et al. 2009 [35]	Iron	Various high-pressure phases	Up to 10	DFT
Morris et al. 2008 [29]	H–Si complexes	Defect structures	32 atom defect cell	DFT
Pickard et al. 2007 [34]	Aluminum hydride	Various high-pressure phases		DFT
Csanyi et al. 2007 [25]	C_6Ca	Carbon sheets	7	DFT
Jiang et al. 2009 [45]	Ta_2N_3	Tetragonal Ta_2N_3	5 and 10	DFT
Lommerse et al. 2000 [43]	Organic compounds	Several		Several
Motherwell et al. 2002 [44]	Organic compounds	Several		Several
Feng et al. 2008 [21]	Li–Be alloys	Four high-pressure phases	Up to 15	DFT

*"One should also mention early papers [Freeman C.M., Newsam J.M., Levine S.M., Catlow C.R.A. (1993). Inorganic crystal structure prediction using simplified potentials and experimental unit cells – application to the polymorphs of titanium dioxide. *J. Mater. Chem.* **3**, 531–535] and [Schmidt M.U., Englert U. (1996). Prediction of crystal structures. *J. Chem. Soc. – Dalton Trans.* **10**, 2077–2082] which pioneered the use of random sampling to crystal structure prediction and defined it in the same forms as it is used now. (Editor's comment)"

In 2003, Bailey et al. studied Ni–Al alloy clusters using the Gupta empirical potential with both random search and genetic algorithm techniques [38]. They generate structures in the same way as in the work described above. Although both methods are successful for small cluster sizes, they find their genetic algorithm to be more efficient than random search, especially in alloy systems which have more degrees of freedom.

Larger systems such as proteins have relatively complicated phase spaces, and prediction of their structure often benefits from specialized techniques [39, 40]. However, Wang et al. found that the efficacy of their genetic algorithm approach was improved by combining it with a random search method. The hybrid method is implemented by replacing some proportion of the worst solutions in each generation by randomly generated ones. This increases genetic diversity in the population and avoids premature convergence to nonglobal minima more effectively than evolutionary mutation operators [41]. The introduction of random structures has played a role in maintaining genetic diversity in other evolutionary algorithms as well, e.g., [42].

Several tests of random searching applied to small organic molecules have been made in the context of a comparison to other methods [43, 44]. The comparisons found random searching (as well as most other methods tested) rather ineffective in predicting the structure of several molecular systems. However, these tests fail to take into account one of the largest benefits of the random search method: simplicity and short amount of researcher time to solution. More importantly, the optimization methods tested made use of a variety of energy codes but were evaluated based on whether or not they found the experimentally known solution. This is problematic since many of the objective functions being optimized may not have even had a global minimum at the physically correct structure. Thus, whether or not a particular optimization method found the experimental structure is a poor indicator of its success. These are interesting works that look at overall strategy but do not separate energy model from optimization scheme in their method evaluation.

The works discussed here are summarized in Table 3.1

3.5
Summary and Conclusions

In summary, we have reviewed the theory of random search techniques and their application to materials' atomic structure prediction in the literature. The method has been successfully applied to many interesting and technologically important systems. For binaries and other systems with relatively small solution spaces, the method is reliable and has the significant advantage of requiring very little programmer time to obtain a solution. More complicated systems may benefit from the use of methods such as genetic algorithms, which incorporate information learned about the system over the course of the algorithm to make better guesses.

References

1. Jansen, M. and Schon, J. C. (2004) Rational development of new materials - putting the cart before the horse? *Nature Materials*, **3**.
2. Jansen, M. and Christian Schon, J. (2006) Design in chemican synthesis - an illusion? *Angew. Chem. Int. Ed.*, **45**.
3. Sholl, D. A. and Steckel, J. A. (2009) *Density Functional Theory: A Practical Introduction*. Wiley, New York.
4. Gale, J. D. and Rohl, A. L. (2003) The general utility lattice program (gulp). *Molecular Simulation*, **29**.
5. Baroni, S., Dal Corso, A., de Gironcoli, S., Giannozzi, P., Cavazzoni, C., Ballabio, G., Scandolo, S., Chiarotti, G., Focher, P., Pasquarello, A., Laasonen, K., Trave, A., Car, R., Marzari, N., and Kokalj, A. Pwscf. http://www.pwscf.org/.
6. Ozolins, V., Wolverton, C., and Zunger, A. (1998) First-principles theory of vibrational effects on the phase stability of cu-au compounds and alloys. *Physical Review B*, **58**.
7. Schon, J. C., Wevers, M. A. C., and Jansen, M. (2003) Entropically stabilized region on the energy landscape of an ionic solid. *Journal of Physics: Condensed Matter*, **15**.
8. Souvatzis, P., Eriksson, O., Katsnelson, M. I., and Rudin, S. P. (2008) Entropy driven stabilization of energetically unstable crystal structures explained from first principles theory. *Phys. Rev. Letters*, **100**.
9. Chaudhuri, I. et. al. (2004) Global optimization of silicon nanoclusters. *Applied Surface Science*, **226**.
10. Solis, F. J. and Wets, R. J.-B. (1981) Minimization by random seearch techniques. *Mathematics of Operations Research*, **6**.
11. Anderson, R. L. (1953) Recent advances in finding best operating conditions. *J. of the American Statistical Association*, **48**.
12. Rastrigin, L. A. (1963) The convergence of the random search method in the extremal control of a many-parameter system. *Authomat. Remote Control*, **24**.
13. Karnopp, D. C. (1963) Random search techniques for optimization problems. *Automatica*, **1**.
14. Spall, J. C. (2003) *Introduction to Stochastic Search and Optimization*. Wiley, New York.
15. Zhigljavsky, A. and Zilinskas, A. (2007) *Stochastic Global Optimization*. Springer, Berlin.
16. Anderssen, R. S. and Bloomfield, P. (2005) Properties of the random search in global optimization. *Journal of Optimziation Theory and Applications*, **16**.
17. Pauschenwein, G. J. (2009) The minimum distance parameterization of crystal lattices. *Journal of Physics A: Mathematical and Theoretical*, **42**.
18. Press, W. H. et. al. (2007) *Numerical Recipes: The Art of Scientific Computing*. Cambridge University Press, 3 edition,
19. Pulay, P. (1980) Convergence acceleration of iterative sequences. the case of scf iteration. *Chemical Physics Letters*, **73**: 393–398,
20. Wales, D. (2003) *Energy landscapes with applications to clusters, biomolecules and glasses*. Cambridge University Press, Cambridge.
21. Feng, J., Hennig, R. G., Ashcroft, N. W., and Hoffman, Roald (2008) Emergent reduction of electronic state dimensionality in dense ordered li-be alloys. *Nature*, **451**.
22. Pickard, C. J. and Needs, R. J. (2006) High-pressure phases of silane. *Physical Review Letters*, **97**.
23. Berry, R. S. (1993) Potential surfaces and dynamics: What clusters tell us. *Chemical Review*, **93**.
24. Stillinger, F. H. (1999) Exponential multiplicity of inherent structures. *Physical Review E*, **59**.
25. Csanyi, G., Pickard, C. J., Simons, B. D., and Needs, R. J. (2007) Graphite intercalation compounds under pressure: A first-principles density functional theory study. *Physical Review B*, **75**.
26. Venkatesh, P. K., Cohen, M. H., Carr, R. W., and Dean, A. M. (1997) Bayesian method for global optimization. *Physical Review E*, **55**.

27. Schon, J. C. and Jansen, M. (1994) Determination of candidate structures for lennard-jones-crystals through cell optimisation. *Ber. Bunsenges. Phys. Chem.*, **98**.
28. Gauzzi, A. et. al. (2008) Maximum T_c at the verge of a simultaneous order-disorder and lattice-softening transition in superconducting cac6. *Physical Review B*, **78**.
29. Morris, A. J., Pickard, C. J., and Needs, R. J. (2008) Hydrogen/silicon complexes in silicon from computational searches. *Physical Review B*, **78**.
30. Pickard, C. J. and Needs, R. J. (2007) Structure of phase III of solid hydrogen. *Nature Physics*, **3**.
31. Pickard, C. J. and Needs, R. J. (2009) High-pressure phases of nitrogen. *Physical Review Letters*, **102**.
32. Pickard, C. J. and Needs, R. J. (2009) Dense low-coordination phases of lithium. *Physical Review Letters*, **102**.
33. Pickard, C. J. and Needs, R. J. (2007) When is H_2O not water? *Journal of Chemical Physics*, **127**.
34. Pickard, C. J. and Needs, R. J. (2007) Metallization of aluminum hydride at high pressures: A first-principles study. *Physical Review B*, **76**.
35. Pickard, C. J. and Needs, R. J. (2009) Stable phases of iron at terapascal pressures. *Journal of Physics: Condensed Matter*, **21**.
36. Pickard, C. J. and Needs, R. J. (2008) Structures at high pressure from random searching. *Physica Status Solidi*, **246**.
37. Lloyd, L. D. and Johnston, R. L. (1998) Modelling aluminum clusters with an empirical many-body potential. *Chemical Physics*, **236**.
38. Bailey, M. S., Wilson, N. T., Roberts, C., and Johnston, R. L. (2003) Structures, stabilities and ordering in ni-al nanoalloy clusters. *European Physical Journal D*, **25**.
39. Scheraga, H. A. (1996) Recent developments in the theory of protein folding: searching for the global energy minimum. *Biophysical Chemistry*, **59**.
40. Wales, D. J. and Scheraga, H. A. (1999) Global optimization of clusters, crystals, and biomolecules. *Science*, **285**.
41. Wang, J., Hou, T., Chen, L., and Xu, X. (1999) Automated docking of peptides and proteins by genetic algorithm. *Chenometrics and Intelligent Laboratory Systems*, **45**.
42. Woodley, S. M., Battle, P. D., Gale, J. D., Richard, C., and Catlow, A. (1999) The prediction of inorganic crystal structures using a genetic algorithm and energy minimisation. *Phys. Chem. Chem. Phys.*, **1**.
43. Lommerse, J. P. M. et. al. (2000) A test of crystal structure prediction of small organic molecules. *Acta Cryst.*, **B56**.
44. Motherwell, W. D. S. et. al. (2002) Crystal structure predition of small organic molecules: a second blind test. *Acta Cryst.*, **B58**.
45. Jiang, C., Lin, Z., and Zhao, Y. (2009) Thermodynamic and mechanical stabilities of tantalum nitride. *Physical Review Leters*, **103**.

4
Predicting Solid Compounds Using Simulated Annealing

J. Christian Schön and Martin Jansen

4.1
Introduction

The poorly developed ability to predict not-yet-synthesized kinetically stable compounds proves to be a major stumbling block on the road to progress in fields ranging from materials science to geo- and astrophysics. This applies in particular to solid-state chemistry since, in contrast to (organic) molecular compounds [1, 2], one even lacks a well-honed heuristic to construct the crystalline structure that a given chemical composition is most likely to exhibit in the solid state [3–8]. Identifying all the feasible (meta)stable periodic atomic configurations, and deciding which among several modifications is the preferred one at a particular temperature and pressure, obviously requires the global exploration of the Born–Oppenheimer surface (commonly called the energy landscape) of the chemical system [7–11].

To achieve this, both the local minima and the barrier structure of the landscape need to be investigated. One should note that it is not sufficient to obtain only the global minimum: all local minima that are surrounded by sufficiently high-energy barriers correspond to metastable modifications that may be of interest regarding their physical and/or chemical properties both in science and in technological applications [7, 8, 12]. Furthermore, there exist many (meta)stable phases that are associated with large locally ergodic regions on the landscape which contain many local minima, e.g., high-temperature modifications with rotating complex anions, or solid-solution phases [11].

Since the beginning of the 1990s, methods for theoretical structure determination and prediction have been developed and applied to a number of systems [13–29]. Since a typical set of, e.g., global exploration runs for the determination of the local minima and other locally ergodic regions can involve millions or even billions of energy evaluations, a modular multiscale approach has become standard [7, 11, 30]. A global search on an empirical energy/cost-function landscape

Modern Methods of Crystal Structure Prediction. Edited by Artem R. Oganov
Copyright © 2011 WILEY-VCH Verlag GmbH & Co. KGaA, Weinheim
ISBN: 978-3-527-40939-6

generates structure candidates[1], employing global exploration techniques such as simulated annealing [35, 36], genetic and evolutionary algorithms [37] (cf. chapter by Lyakhov and Oganov), the threshold or lid algorithm [38–40], the metadynamics approach [41] (cf. chapter by Martonak), or the ergodicity search algorithm (ESA) [42]. These candidates are subsequently locally optimized on full quantum mechanical level using, e.g., the Hartree–Fock approximation or density functional theory [20, 43, 44].

In this chapter, we give an overview over the application of simulated annealing and closely related stochastic global exploration algorithms to the structure prediction of solid compounds. First, we discuss the concept of locally ergodic regions on the energy landscape of chemical systems. Next, simulated annealing and related random walker methods used for the study of energy landscapes are described in some detail. The third part of this chapter gives an overview of structure prediction using these algorithms, with examples ranging from predictions of simple binary and ternary solids, molecular crystals, zeolites and phase diagrams, to the determination of structures by combining experimental information and potential energies. A critical evaluation of the state of the field concludes this work.

4.2
Locally Ergodic Regions on the Energy Landscape of Chemical Systems

A general detailed introduction to multiminima energy landscapes has been given in the chapter by Wales. Here, we address only one aspect of such landscapes that is crucial to the issue of predicting (meta)stable solid compounds and phases: the concept of locally ergodic regions. Starting point for any prediction of not-yet-synthesized compounds up to the derivation of phase diagrams without recourse to experimental information is the Born–Oppenheimer surface over the $3N$-dimensional space of all atom arrangements, the so-called configuration space of the system. This energy hypersurface is commonly denoted as the energy landscape of the chemical system [7–9, 11, 12, 31, 45–47], and the dynamics of the system corresponds to a trajectory on the landscape.

The crucial step in going from the classical mechanical description above to the thermodynamic one is the determination of the so-called locally ergodic regions on the energy landscape [9]. For a given temperature T, a subset $\mathcal{R}$ of the configuration space is called locally ergodic on the observation time scale t_{obs}, if the time $\tau_{eq}(\mathcal{R}; T)$ it takes for the system to equilibrate within $\mathcal{R}$ is much shorter than t_{obs}, while the

---

1) Here, different levels of approximation and subsequent refinements for the description of the energy landscape are applied, both with regard to the choice of energy function (e.g., *ab initio* energies [27, 29, 31, 32], empirical atomic interaction potentials [6, 7, 9], or atom-group based cost functions [22, 24, 33]) and concerning the quantities that describe the atom configuration and are all varied during the global optimization (such as single atoms, groups of atoms [24, 33], or nodes in bond networks [34]). For a review of the current state of the field of structure prediction using only *ab initio* energy functions, see Ref. [31].

time $\tau_{\text{esc}}(\mathcal{R}; T)$ it takes for the system to leave the region $\mathcal{R}$, the so-called escape time, is much larger than t_{obs},

$$\tau_{\text{esc}}(\mathcal{R}; T) \gg t_{\text{obs}} \gg \tau_{\text{eq}}(\mathcal{R}; T) \tag{4.1}$$

If this holds true, then the ergodic theorem tells us that we can replace the time averages of observables $O(\vec{R}(t), \vec{P}(t))$ along a trajectory of length $t_{\text{obs}} = t_2 - t_1$

$$\langle O \rangle_{t_{\text{obs}}} = \frac{1}{t_{\text{obs}}} \int_{t_1}^{t_2} O(\vec{R}(t'), \vec{P}(t')) dt' \tag{4.2}$$

inside the locally ergodic region $\mathcal{R}$ by the (Boltzmann) ensemble average of this observable

$$\langle O \rangle_{\text{ens}}(T) = \frac{\int O(\vec{P}, \vec{R}) \exp(-E(\vec{P}, \vec{R})/k_B T) d\vec{P} d\vec{R}}{\int \exp(-E(\vec{P}, \vec{R})/k_B T) d\vec{P} d\vec{R}} \tag{4.3}$$

restricted to the region $\mathcal{R}$,

$$|\langle O \rangle_{t_{\text{obs}}} - \langle O \rangle_{\text{ens}}(T)| < a \tag{4.4}$$

Of course, this "equality" holds only within an accuracy a, since only local and not global ergodicity is asserted. In particular, we can compute for every locally ergodic region $\mathcal{R}_i$ the local free energy

$$F(\mathcal{R}_i, T) = -k_B T \ln Z(\mathcal{R}_i, T) = -k_B T \ln \sum_{j \in \mathcal{R}_i} \exp(-E(j)/k_B T) \tag{4.5}$$

and thus apply the usual laws of thermodynamics to the system as long as it remains within the region $\mathcal{R}_i$.

As a result, for any given observation time scale t_{obs} and temperature T, the configuration space of the chemical system is split into a large number of disjoint locally ergodic regions, with the remainder of the configuration space consisting of transition regions connecting the locally ergodic regions. Each such region corresponds to a kinetically stable compound of the chemical system on the time scale of observation.[2] At low temperatures, the regions with the lowest free energy are basins around individual local minima that correspond to crystalline modifications of the system, while structures containing defects are also associated with local minima but with higher energies. At elevated temperatures and on sufficiently long time scales, locally ergodic regions will typically encompass many local minima, e.g., the region may consist of a large basin containing both the

---

[2] Note that if the system has been given an essentially infinite time $\tau_{\text{eq}}^{\text{global}} \gg \tau_{\text{esc}}(\mathcal{R}_i)$ to equilibrate before we perform our measurement on the timescale $t_{\text{obs}} (\ll \tau_{\text{esc}}(\mathcal{R}_i))$, the system can be treated as globally ergodic, and the likelihood of finding the system at the time of the measurement in a particular locally ergodic region $\mathcal{R}_i$ is given by $p(\mathcal{R}_i)$. As a consequence, the locally ergodic region with the lowest free energy has the highest probability of being occupied during the measurement, and the compound corresponding to this region is customarily designated to be the thermodynamically stable phase.

perfect crystalline minimum and the minima corresponding to equilibrium defects of this structure.

In principle, the determination of locally ergodic regions consists of three steps: the generation of a candidate for such a region, the verification that the candidate is locally equilibrated on the time scale of observation, and the verification that the candidate is kinetically stable on the time scale of observation [48]. One should note that while many methods have been developed to generate structure candidates (for an overview see, e.g., Ref. [11] and the various chapters in this book), the remaining two steps of verifying that these candidates are equilibrated and kinetically stable on the relevant observational time scales are in most instances reduced to only checking whether the candidate corresponds to a local minimum of the energy. The reason for this is that further investigations such as the measurement of the flow of probability on the landscape and the determination of the energetic and entropic barriers surrounding the locally ergodic regions [49] are still far from trivial and require a very large computational effort, especially if one attempts to compute these barriers on the *ab initio* quantum mechanical level.

This restricted focus on individual local minima turns into a serious problem if one wants to determine locally ergodic regions at elevated temperatures. Four types of locally ergodic regions that are important at high temperatures but are difficult to identify without recourse to experimental data need to be distinguished: (i) high-lying local minima that are competitive with the global minimum at high temperatures due to, e.g., a soft vibrational spectrum, (ii) entropically stabilized regions, and regions exhibiting (iii) local or (iv) global controlled disorder resulting in a configurational entropy contribution to the free energy of the region.[3]

Case (i) is the most straightforward one to address – here one "only" needs to make sure that one determines all relevant local minima of the energy landscape and computes the local free energy of all locally ergodic regions with sufficient accuracy. Case (ii) refers to those locally ergodic regions that are not associated with deep local minima (or contain no minima at all!), but which are stabilized by the fact that the exit from such a region takes place on much larger time scales than the time needed to locally equilibrate within the region [49].[4]

Cases (iii) and (iv) apply to locally ergodic regions that contain many local minima that can equilibrate to a certain degree on the relevant time scales (of synthesis and/or measurement). Typical examples of local controlled disorder are

---

3) We employ the term "controlled disorder" for those systems where some underlying long-range structural order exists on top of which the disorder manifests itself. The underlying order is typically a periodic arrangement of atoms or groups of atoms, irrespective of their actual type and/or electronic state, and/or orientation or displacement. This distinguishes such compounds from the structural glasses and amorphous solids without long-range translational order. If the degree of freedom associated with the disorder is frozen out at low temperature, we are dealing with, e.g., orientational glasses (frozen rotation), chemically disordered systems (frozen diffusional atom exchange), or spin glasses (frozen spin configuration).

4) The major problem is to identify such a region, since often the local minima involved have so small barriers that the global optimization algorithms will pass them by; an algorithm that can detect these regions is, e.g., the ESA [42].

compounds with (independent) rotating complex ions, where the centers-of-mass of these rotating atom groups form a stable lattice, and the multi-atom-group rotation consists of jumps between many close local minima on the energy landscape. In contrast, global controlled disorder is found in solid solutions or compounds with partially occupied atom positions. Here, each particular arrangement of the atoms on a common stable sublattice constitutes a local minimum on the energy landscape, and the multiplicity of such (energetically approximately equal) arrangements yields the configurational entropy contribution to the free energy. In contrast to the local controlled disorder where the individual atom groups rotate essentially independently of each other, the controlled disorder in an alloy manifests itself in a global rearrangement by an exchange of atoms of different types. For a more detailed discussion of locally ergodic regions and their relationship to solid compounds, we refer the reader to Refs. [11, 31].

As a consequence, the central quantities of interest for the purpose of structure prediction are not only the local minima and saddle points, but also the locally ergodic regions [9] and transition regions [50], the local densities of states, and the flow of probability on the landscape with the corresponding barrier landscape consisting of (generalized) barriers [49] such as energetic, entropic, and kinetic barriers [51]. In practice, determining the relevant local minima has proven to be the cornerstone of essentially all methodologies for unbiased structure prediction. In the following section, we will therefore describe in detail the simulated annealing algorithm and closely related global search techniques for finding local minima, and only in passing address the issue of investigating the barrier structure of the landscape; gradient-based techniques, genetic and evolutionary algorithms, exhaustive searches, extremal optimization, etc., some of which are discussed in detail in other chapters, will not be included. Similarly, we refer the reader to the literature [11] for an overview and further references regarding the analysis of the barrier structure, the determination of local densities of states, and the computation of the local free energies.

4.3
Simulated Annealing and Related Stochastic Walker-Based Algorithms

4.3.1
Basic Simulated Annealing

Algorithms that employ random walkers to explore energy landscapes are based on the following four fundamental features: (1) a set of random walkers that can be noninteracting, interacting, and/or learning from each other. (2) A configuration (or solution) space $S = \{\vec{x}\}$ together with an energy (or cost) function $E(\vec{x})$ that can stay unchanged or evolve as the algorithm proceeds. (3) A moveclass (or neighborhood) $\mathcal{N}(\vec{x})$ which gives for each state $\vec{x}$ the neighboring states that can be accessed with a certain probability by the random walker if it is at state $\vec{x}$. This moveclass can remain unchanged or evolve as the algorithm proceeds. (4) An acceptance criterion according to which the walker makes the move to the neighbor state selected. Again, this criterion can (and often does) vary during the run.

The prototype of such an algorithm is the so-called Monte Carlo Metropolis algorithm describing a single walker at a constant temperature T, which employs the Metropolis acceptance criterion. At the beginning, a starting point $\vec{x}_0$ for the walker is chosen, either at random or according to some deterministic scheme. The move from the ith to the $(i+1)$th position of the walker along this trajectory takes place as follows: from the neighborhood $\mathcal{N}(\vec{x}_i)$ of the current state $\vec{x}_i$ we select, at random with probability according to the moveclass, a target state $\vec{x}_{\text{target}} \in \mathcal{N}(\vec{x}_i)$. Next, we compute the difference in energy between these two states, $E(\vec{x}_{\text{target}}) - E(\vec{x}_i)$. If $E_{\text{target}} \leq E_i$, the move is accepted. If $E_{\text{target}} > E_i$, a random number $0 \leq r \leq 1$ is generated. If now

$$\exp(-(E_{\text{target}} - E_i)/T) \geq r \tag{4.6}$$

then the move is accepted, i.e., $\vec{x}_{i+1} = \vec{x}_{\text{target}}$. Else, the walker stays at $\vec{x}_i$, i.e., $\vec{x}_{i+1} = \vec{x}_i$. This procedure is repeated until the maximal number of steps $N_{\max}$ has been performed, and the full trajectory $\{\vec{x}_0, \vec{x}_1, \ldots, \vec{x}_{N_{\max}}\}$ has been obtained.

Under certain conditions, the trajectory covers the space $\mathcal{S}$ ergodically such that the time average of some quantity $Q(\vec{x})$ over the trajectory equals the Boltzmann ensemble average

$$\frac{\sum_{\vec{x}} \exp\left[-E(\vec{x})/T\right] Q(\vec{x})}{\sum_{\vec{x}} \exp\left[-E(\vec{x})/T\right]} \approx \frac{1}{N_{\max}} \sum_{i=0}^{N_{\max}} Q(\vec{x}_i) \tag{4.7}$$

In reality, for complex multiminima systems, $N_{\max}$ is usually much too short to yield global ergodicity, especially if T is varied or the landscape evolves during the run. But the heuristic picture of nearly ergodic behavior, i.e., of the walker along its trajectory being approximately in (local) thermal equilibrium, can often be useful in analyzing the behavior of such algorithms [52–54].

The most common generalization of the Metropolis algorithm consists in varying the temperature during the run. The so-called simulated annealing algorithm [35, 36] works by slowly lowering the temperature, thus moving the walker to (on average) states with lower and lower energy. The expectation is that if one proceeds slowly enough, the walker will at the end of the run have reached the global minimum of the energy landscape. Another well-known algorithm is the stochastic quench, where one performs a Metropolis random walk at zero temperature, i.e., only steps that lower the energy are being accepted until a local minimum has been reached.[5]

That leads directly to the most straightforward approach to global optimization, i.e., multiple (stochastic) quenches, possibly combined with a gradient-based

---

5) Especially if one performs such stochastic quenches from high-lying states one can gain additional information beyond that obtained by performing a gradient minimization, because by repeating the quench from the same starting point for different random number sequences, one can determine whether the starting point is associated with only one or several local minima, i.e., whether it "belongs" to a single basin or a transition region, respectively [50].

minimization. This local minimization procedure is performed for a very large number of starting points that are generated either by systematically or randomly scanning the configuration space [55], by chemically inspired choices [56], e.g., via network model generation [25, 57, 58] (cf. chapter by Blatov and Proserpio) or selection of structures from databases [59, 60], or by performing long Monte Carlo simulations at constant temperature, where periodically quenches are performed along the trajectory of the system [61, 62]. For more details on so-called random search algorithms, see the chapter by Tipton and Hennig.

Figure 4.1 shows a typical simulated annealing run for a complex multiminima energy landscape (the system was a silicate consisting of 26 atoms/repeated cell, belonging to five different atomic species). Along the trajectory of the walker, many quenches into local minima were performed. One notes that the energies of the local minima that are encountered by the random walker decrease as the walker reaches deeper multiminima basins on the landscape. This demonstrates the reason why simulated annealing was originally introduced to deal with (NP-)hard global optimization problems: For such large problems, the random search methods that had been employed earlier could only reach the higher lying minima like those seen in Figure 4.1 at the beginning of the simulated annealing run (when the

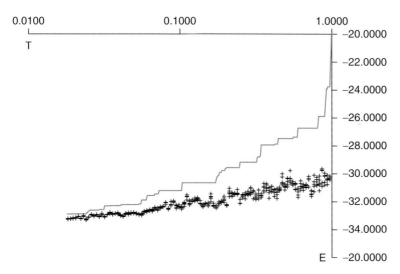

Figure 4.1 Best-energy-seen-so-far of a random walker during a typical simulated annealing run for a 26-atom system (solid line), and the energies of 500 local minima seen along the trajectory (plusses), both as a function of temperature. The temperature scale along the x-axis is logarithmic (in eV), and the energy is in eV/atom along the y-axis. Since $T(n) \propto f^n (T(0) = 1.0)$, with $f < 1$, the simulation time of the algorithm increases linearly toward lower temperatures in this plot. Note that the minima encountered at the beginning of the run when the trajectory of the walker was still essentially random (i.e., the set of minima generated would be similar to what one would expect from a random search) were all higher than the energy found at the end of the simulated annealing run, even before the final quench.

temperature is still quite high and thus the walker essentially performs a nearly free random walk on the landscapes with most moves being accepted). Unless one can design efficient heuristics to generate good (random) starting points for the many local minimizations, the random search algorithm remains stuck in the high-lying part of the landscape. In contrast, the walker in simulated annealing digs deeper and deeper into the landscape until it reaches those basins where the global minimum resides. A quantitative comparison of stochastic quenches with simulated annealing runs for the case of a small Lennard–Jones system (eight atoms/cell) can be found in Ref. [16]. Both random search and simulated annealing were successful in reaching the global minimum, but the percentage of times the low-energy structures were found was considerably higher in the case of simulated annealing.

4.3.2
Adjustable Features in Simulated Annealing

4.3.2.1 Choice of Moveclass

When using simulated annealing to identify possible crystalline modifications in a chemical system, the most basic moves are the displacement of atoms or groups of atoms (that can be predefined as building units or selected ad hoc) and the change of the periodically repeated simulation cell. Furthermore, a very powerful move is the exchange of atoms or groups of atoms [7, 18]. For standard simulated annealing, the ratio between the number of atom displacement moves and moves where the cell is varied and atoms are exchanged, usually lies between 1/2 and 1/3 of the number of atoms. The optimal value of this ratio depends on the system under investigation, of course, e.g., taking into account the size and shape of various multiatom building units. The selection of an appropriate moveclass is of great importance for the efficiency of the global search. This dependence of the results on the moveclass was demonstrated in the eight-atom Lennard–Jones system study mentioned above [16].

Regarding the total number of atoms involved, while in many systems the most important structures can already be found for two formula units/simulation cell ($z = 2$), one should always include higher values of z, i.e., perform global searches for at least $z = 2$, 3, and 4; preferably going up to $z = 8$. Of course, if one formula unit already involves 20+ atoms, going beyond $z = 2$ quickly reaches the limits of the simulated annealing method for calculations on a single processor.

In addition, one can employ moves that change the composition of the set of atoms present in the simulation cell; however, in this case, one needs to take the chemical potential into account, i.e., the price one pays for adding/removing an atom [7, 63]. Typically, one uses either the cohesive energy/atom in the element (at the current temperature of the annealing process) or establishes a fixed vapor pressure in order to represent the chemical potential. In practice, it has proven to be more efficient in most situations to repeat the global search for many different compositions instead of varying the composition during an individual simulated annealing run. Other quantities that can be changed during the search are the ionization states of the atoms, i.e., both the charge that enters the Coulomb term and the corresponding ionic radius in the dispersion and repulsion terms of the potential are varied during the run. Here, one adds the ionization energy or electron affinity for cations or anions, respectively, to the total energy [7, 18].

In general, one can perform simulated annealing both via discrete-step walkers and via, e.g., constant temperature molecular dynamics simulations. In the latter case, one reduces the temperature of the system, i.e., the average kinetic energy of the particles, during the simulation.[6] Quite generally, a great advantage of the discrete steps in Monte Carlo simulations for global optimization purposes is the large freedom in choosing a moveclass most appropriate to the type of exploration being performed. In contrast to, e.g., molecular-dynamics-based algorithms, one can replace the physically realistic moveclass of moving one or a few atoms by a small amount by a more optimization effective moveclass that allows larger changes in the atom configuration during each move, in order to explore a larger part of the landscape during, e.g., a global optimization run. In this case, it is sometimes efficient, to combine such large moves every time with a quench and/or a gradient-based minimization. Then one applies the acceptance criterion to the minimum configurations; this scheme is often called basin hopping [65, 66] (for more details, cf. e.g., the chapters by Goedecker and Wales).

However, one should also recall that one can, in principle (although in practice often only a posteriori), remove any local minimum except the global one by designing a special moveclass such that the walkers "tunnel" from the bottom of a minimum basin into a deeper basin. As long as the goal is only to find the global minimum this does not constitute a problem and would even be helpful. But in the case of structure prediction, we are usually interested in all locally ergodic regions of the landscape, corresponding to kinetically stable modifications. Thus one needs to be careful about the choice of moveclass, in order not to eliminate important minima.[7] As a consequence, in spite of its disadvantages as a global search mechanism, the "natural" moveclass implied in the Newtonian dynamics can in some situations appear to be the more appropriate one to use during the global search.

Another question is how one can deal with systems that contain not only continuous but also discrete degrees of freedom, such as spin structures, that might noticeably contribute to the overall energy via some semiempirical interaction potential (unless they are incorporated directly in an *ab initio* calculation and thus are integrated out during the energy evaluation). Since the spins take on discrete values, they are more easily dealt with within the framework of discrete random

---

6) When studying problems derived from physics and chemistry, one often encounters a certain prejudice in favor of the molecular dynamics instead of Monte Carlo-based methods, since the former appear to be more "realistic" than the random walker algorithms. However, once the simulation times are longer than the typical vibrational time scales, and one employs a moveclass consisting of local atom displacement moves, the expectation values of thermodynamic quantities are approximately the same for both dynamics [64]. This is particularly relevant if one wants to derive local densities of states from the trajectories of the walkers.

7) Note that this caveat does not only apply to random walker algorithms but also to, e.g., genetic algorithms.

walkers. A similar issue arises when one attempts to reduce the number of the degrees of freedom in the system by constructing rigid or floppy building units. Again, a Newtonian-type dynamics requires considerably more effort to implement even for primary building units; and in the case of secondary building units, the very artificial interaction potentials associated with atom-mergers greatly reduce the realism of the corresponding molecular dynamics simulation.

4.3.2.2 Temperature Schedule and Acceptance Criterion

Besides the moveclass, there are a number of other features that can be adjusted to increase the efficiency of the algorithm [54, 67]. The temperature schedule $T(n)$, where n counts the number of moves along the trajectory, can be optimized; common schedules consist in an exponential or linear decrease of temperature with n. A detailed analysis of the influence of the parameters in the temperature schedule on the outcome the annealing has been performed in Ref. [16] for an eight-atom Lennard–Jones system.

Also quite popular are schedules involving temperature cycling [68, 69] where the temperature periodically increases and then decreases again, and adaptive schedules [70, 71] that take properties of the landscape explored up to now into account. One class of adaptive schedules is based on the frequency of accepted moves during the most recent time window, where one adjusts the temperature and/or the moveclass (e.g., the size of the random atom displacements) such that an acceptance rate of about 50% is achieved.[8] Other approaches derive from finite-time thermodynamics concepts, where the speed at which the temperature is decreased is taken to be inversely proportional to the fluctuations in energy along the walker's trajectory [70, 71]. A simplified version of this approach employs, as criterion for lowering the temperature, the fact that the average energy during a time window along the trajectory no longer decreases but increases (possibly by an amount larger than the fluctuations in energy during the previous time window) [72].

Finally, the acceptance criterion can be modified; the most popular ones accept a move according to the classical Metropolis distribution [73], a Fermi-function-like distribution (the so-called fast-annealing [74]) or the Tsallis distribution [75, 76], or based on a temperature-dependent acceptance threshold [77]. The Tsallis distribution replaces the exponential in Eq. (6) by a power law, while in threshold accepting, one accepts the move as long as $E(\vec{x}_{\text{target}}) - E(\vec{x}_i) < T$, i.e.,

$$\exp(-(E_{\text{target}} - E_i)/T) \geq 1/e \tag{4.8}$$

In this way, threshold accepting dispenses with generating a random number and computing the exponential, thus speeding up the calculation. The price one pays is the exclusion of unusual moves to target states that are associated with large energy differences, and the loss of thermal equilibrium in the Boltzmann

---

8) Keeping the temperature constant and modifying only the moveclass is quite popular in hybrid methods, such as basin hopping, that combine local minimizations and simulated annealing.

ensemble. Of course, in many cases these states can be reached by climbing a "ladder" of intermediate states that are separated by energy differences smaller than T. Finding this sequence of states during random neighbor selections is rather unlikely, however. On the other hand, this is balanced, when using the standard Metropolis criterion, by the low probability of drawing a random number smaller than $\exp(-(E_{\text{target}} - E_i)/T)$.

This is an example of the subtle trade-offs one has to make when designing a stochastic exploration or optimization algorithm. Another such trade-off appears in the choice of the moveclass. On the one hand, one can try to select or construct neighborhoods $\mathcal{N}(\vec{x})$ such that the energy barriers between, e.g., the minima are rather low. In this case, the algorithm never gets stuck in high-lying local minima: there is always a low saddle available for exiting a minimum. On the other hand, this approach typically results in very large and complicated neighborhoods, and as a consequence the progress of the algorithm, measured, e.g., in $\langle E(\vec{x}) \rangle_{t_{\text{window}}}(t)$, is mainly determined by the ability to find this exit route by essentially a random search. Instead of a high energetic barrier, we are now dealing with an entropic or kinetic barrier [49]. Thus, optimizing a stochastic minimization algorithm is very difficult to do in a generic fashion – instead one will want to incorporate as much a priori information about the landscape as possible.

4.3.2.3 Extensions and Generalizations of Simulated Annealing

Incorporating landscape information into the algorithm can be achieved in a sequential or parallel fashion. The first way is used in, e.g., taboo-like searches [78–80], where those regions or minima basins that have already been visited by the walker are excluded from the remainder of the search (or for the next search). Typically, this is achieved either by rigid exclusion constraints or via penalty terms added to the energy function [81]. A crucial issue here is the length of the memory chain; combining a taboo search with quenches and large moves like in basin hopping schemes can alleviate the memory problem to a certain extent [26].

Trying to collect and exploit landscape information instantaneously has been the motivation behind the many multiwalker implementations of simulated annealing [67, 82], such as the Demon algorithm [83] where walkers are moved from high-energy states to already discovered low-energy regions,[9] and methods that generate an averaged landscape [84–87] such as conformation-family Monte Carlo [88] where several configuration families each consisting of similar structures compete, superposition state molecular dynamics [89] and SWARM molecular dynamics [90]. Other multiwalker methods are multioverlap dynamics [91–93], parallel tempering [94], and J-walking [95] where different walkers run at different temperatures and periodically switch positions (or temperatures), in order to overcome barriers more efficiently.

---

9) This algorithm needs to be finely tuned since it can easily happen that too many walkers are placed into one basin at too early a stage, leading to a loss in diversity and the algorithm being stuck at a mid-level minimum; in such a case, the algorithm behaves like a more efficient multiple quench.

The methods mentioned above aim in some way at addressing global topological features of the landscape by deleting whole regions of the landscape or by exploring the landscape on many different time and/or length scales in parallel. In a somewhat different way, landscape modifications and/or simplifications that do not change the topology but aim for a more efficient barrier crossing are employed by algorithms [81, 96–104] that proceed, e.g., by locally elevating visited areas [81], by lowering barriers relative to the local minima [100], by stochastic tunneling [101, 105–107], by dynamic-lattice searching [108], or by modifying the potential between the atoms [98]. Typically, the landscape modifications during such accelerated Monte Carlo/molecular dynamics runs are adaptive, i.e., they vary with the progress of the simulation and depend on the information already gained about the system, in this way being related to the taboo-searches. Due to the plethora of different methods, we refer the reader to the cited literature for more details.

A rather different line of attack is taken by the lid-based methods for continuous energy landscapes. Examples are the deluge algorithm [109], where an energy lid that must not be crossed during the random walk is slowly lowered from very high lid values[10], and the threshold algorithm [40] (originally developed as an implementation of the lid algorithm [38, 110, 111] for continuous landscapes), where for a sequence of energy lids the walker is allowed to move below the lid with every move accepted as in the deluge algorithm, and one checks periodically whether new local minima have been reached by performing several quench runs from stopping points along the trajectories below the current energy lid.

Finally, as discussed in the previous section, determining the locally ergodic regions at low temperatures corresponds to identifying the local minima on the energy landscape and checking that they are surrounded by sufficiently high-energy barriers [40, 112]. One should keep in mind that we are not only interested in the global minimum – all minima with low energies and sufficiently high barriers surrounding them are of importance since they represent potentially valuable metastable compounds.

However, the situation is more complex when one tries to identify locally ergodic regions at higher temperatures. In principle, one can use long (Monte Carlo) simulations and attempt to visually identify stable structures about which the system oscillates, even if the structure does not correspond to a single, or even any, local minimum of the energy landscape [113–115]. A more systematic approach is the so-called ESA [42], where one registers the fluctuation of indicator variables, for instance the potential energy or the radial distribution function, within time windows during the long (Monte Carlo) simulations. If the average value of these variables jumps between two windows by more than the fluctuation, this suggests the existence of a new locally ergodic region. Next, swarms of short (Monte Carlo)

---

10) The name "deluge algorithm" comes from the equivalent problem of maximizing the function $-E(\vec{x})$, where the decreasing energy lid is transformed to a rising water level (the deluge) driving the walker to the peaks of the landscape.

simulations starting from points along the trajectory in the time window are employed to verify whether the system is in local equilibrium in this region, and long simulations for a number of temperatures are used to measure the probability flow from the region and thus the escape time. Unsurprisingly, searching for locally ergodic regions in this fashion is quite expensive computationally.

Clearly, the concern about the stability of the locally ergodic regions leads to the study of the barrier structure of the landscape. Quite generally, the procedures to analyze this structure, e.g. to find saddle points and transition paths and to measure the probability flow on the landscape, are considerably more involved and often less robust than the methods employed to determine local minima. Most of the methods used to gain information about the barriers separating two local minima, or about the transition regions connecting the various locally ergodic regions, do not rely on stochastic explorations. Instead, they are essentially deterministic schemes based on information about the local minima, the gradients, and the curvatures (via the eigenvectors of the Hessian at the minima or saddle points) [116–141], although some stochastic procedures such as the threshold algorithm are also employed [39, 40, 112, 50]. For more details, we refer to the literature, e.g., [11].

4.4
Examples

Since the topic of this chapter is structure prediction using simulated annealing-type methods, all examples presented below employ stochastic single- and multiwalker algorithms as an essential feature. Moreover, we exclude those random-walker-based procedures that are discussed elsewhere in this book, such as basing hopping or random searches. Furthermore, after two decades of research, there exist very many examples of structure prediction using simulated annealing; due to lack of space, only a few, all dealing with extended solids, can be presented in some detail here.[11]

In order to organize the presentation in a sensible fashion, we note that the studies one finds in the literature under the heading of "structure prediction" of solids can be divided into three different classes: On the one extreme is the structure determination, where important structural information, typically a unit cell and its content, is known from the experiment. On the other extreme is the unrestricted structure prediction, where only the stoichiometry but neither the unit cell nor the number of formula units/(primitive) unit cell is known. And if we

---

11) Large finite molecules such as clusters [142–152], polymers [9, 153–155] or proteins [156–162] constitute a different class of chemical systems where over the past two decades much effort has been invested in the identification of their structures, often using simulated annealing as the global optimization technique. Since these systems contain a finite number of atoms and lack the complicating feature of periodic boundary conditions, they have served as a testing ground not only for simulated annealing but for many of the global optimization and structure prediction techniques mentioned in this book. As this chapter deals with structure prediction of solids, we refer to the literature mentioned for more details and specific examples.

only know the general chemical system, i.e., the composition is not fixed either, structure prediction becomes equivalent to the prediction of the phase diagram of the system, at least at low temperatures.

Between these two extremes lies the case of restricted structure prediction, e.g., the prediction of structures in systems where certain structural elements or local environments of atoms are predefined or assumed at the outset, such as primary and secondary building units. Here, one employs some general chemical constraints that are quite plausible in the system under investigation but are, strictly speaking, not admissible in a true structure prediction. The use of such information to simplify the prediction problem can be justified by noting that these units are given at the beginning of the synthesis and will not change throughout (e.g., during the formation of molecular crystals), or that the particular synthesis route will preform such units before the actual crystallization takes place (e.g., during the synthesis of polyoxometalates), or that the system by itself always (?!) establishes these units in the final modifications observed (e.g., typical coordination polyhedra such as SiO_4-tetrahedra in silicates), or, finally, by noting that in some practical applications one is only interested in those modifications that obey these constraints and thus only focusses on the possible existence and stability of these compounds (e.g., when studying electronic properties of compounds that exhibit octahedral oxygen coordination polyhedra of various transition metal ions).

4.4.1
Structure Prediction

4.4.1.1 Alkali Metal Halides

One of the first ionic systems whose energy landscape has been investigated [18] in detail using simulated annealing without recourse to experimental data is NaCl. By varying atom positions, cell parameters, and atomic charges, a large number of local minima was found on the empirical energy landscape, and the global minimum of the landscape corresponded to the experimentally observed rock salt structure. For the most part, four formula units/simulation cell ($z = 4$) were employed in this study; however, simulated annealing runs with up to 40 atoms/simulation cell ($z = 20$) have been performed for this system.

The structures of most of the energetically low-lying minima could be identified with typical AB-structure types like NiAs, PtS, CsCl, or sphalerite. However, one deep-lying local minimum, denoted $Na^{[5]}Cl^{[5]}$ (the so-called 5–5-structure type), exhibited a structure type previously unknown in ionic systems. Here, Na^+ and Cl^- were coordinated by Cl^- and Na^+, respectively, in a trigonally bipyramidal fashion, resulting in a topology that resembled the one of hexagonal BN.

The energy barrier stabilizing this structure was only moderately high (corresponding to ≈ 100 K), suggesting that this modification might be difficult to synthesize with traditional solid-state synthesis methods in the NaCl system. Thus it came as a pleasant surprise when this new predicted structure type was found experimentally [163] as the aristotype of Li_4SeO_5, where Li and Se occupy the Na positions and O the Cl positions in the $Na^{[5]}Cl^{[5]}$ structure, respectively. By now, this structure type has also been observed during the growth of ZnO films [164].

Analogous global optimizations have been performed since for all twenty alkali halides [165] for a wide range of pressures. Similar to the case of NaCl, many possible modifications were found that included both well-known AB-structure types and previously unknown structures. Finally, simulated annealing was employed to find the minima of LiF on the *ab initio* energy landscape at standard pressure, where both the Hartree–Fock approximation and density functionals were used to compute the energy [29]. The same relevant minima as with the empirical potential were obtained, including the rocksalt-, the wurtzite- and sphalerite-, the NiAs-, and the 5–5-structure tpye. This study served both as a valuable validation of the many landscape explorations based on empirical potentials and as a proof-of-principle for the feasibility of global stochastic explorations on *ab initio* energy surfaces.

In this context, one should mention two important experimental substantiations of the energy landscape approach to structure prediction; the successful synthesis of LiBr [166] and LiCl [167] in the predicted [165] wurtzite structure type. Using the low-temperature atom beam deposition method (LT-ABD) [168], it proved possible to synthesize these new metastable modifications reproducibly for a well-defined set of synthesis parameters.

4.4.1.2 Na_3N

Another deceptively simple chemical system where a successful synthesis followed the prediction is Na_3N.[12] In several studies [170–172], the enthalpy landscapes of all alkali nitrides M_3N (M = Li, Na, K, Rb, Cs) were explored with simulated annealing and the threshold algorithm for a wide range of pressures. This resulted in a large number of structure candidates, including e.g., Li_3N-, Li_3P-, Li_3Bi-, $AuCu_3$-, Al_3Ti-, ReO_3-, and UO_3-structure type, plus many previously unknown structure types. Figure 4.2 shows a part of the tree graph for the empirical-energy landscape of Na_3N containing some of the most important local minima. Ab initio calculations using the Hartree–Fock approximation suggested that for Na_3N, the most likely candidate would be the Li_3P-type, followed by the Li_3N- and the ReO_3-type, with the Li_3Bi-type expected at high pressures.

Using the LT-ABD method [168], Na_3N was subsequently synthesized several years ago [173]. While the primary product as obtained from the activated elements adopted the ReO_3-type of structure, recent high-pressure experiments resulted in its transformation into Li_3N-, Li_3P-, and Li_3Bi-structure types [174, 175]. Furthermore, at intermediary pressures another modification exhibiting the YF_3-type was observed [174, 175] that resembles several structures found as local minima on the enthalpy landscapes of the alkali nitrides.

---

12) The inability to synthesize any compound of this composition had for many decades been hailed as a blatant violation of the homologue rule, since Li_3N can be synthesized directly from the elements at ambient conditions [169].

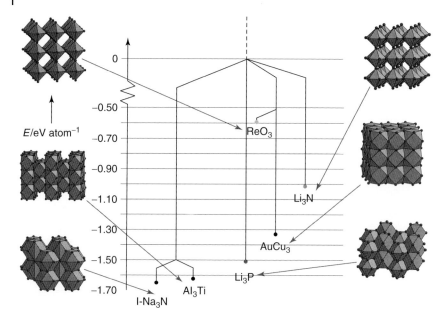

Figure 4.2 Excerpt of the tree graph of the energy landscape of Na_3N on empirical energy level at standard pressure, depicting some of the most important local minima (170–173). I-Na_3N corresponds to a strongly distorted Li_3Bi-structure type with 12(+2)-fold coordination of the nitrogen atoms by sodium atoms.

4.4.1.3 Mg(BH$_4$)$_2$

The modular approach described above, i.e., global exploration with simulated annealing of the energy landscape using a simple empirical potential followed by local optimization on *ab initio* level, was also employed in the study of the ternary system $Mg(BH_4)_2$ [176] that has been discussed as a material for hydrogen storage purposes.[13] Ozolins *et al.* compared the results obtained through this approach with two alternative methods to generate structure candidates: (a) selecting candidates from related structures in the ICSD database and minimizing their energy, and (b) performing *ab initio* molecular dynamics simulations starting from the best structures found via the database search. They observe that the global search on the empirical landscape produces additional structure candidates, one of which appears to be lower in energy than the ones produced by the other two methods, underlining the importance of performing unbiased global searches as part of the structure prediction.

13) The authors actually created a new name, prototype electrostatic ground state search (PEGS), for their procedure, in the belief that this was a new method – a classical case of reinvention in science.

4.4.1.4 Elusive Alkali Metal Orthocarbonates Balancing $M_4(CO_4)$ and $M_2O + M_2(CO_3)$, with M = Li, Na, K, Rb, Cs

The existence of esters of the hypothetic acid $H_4(CO_4)$ suggests that its salts, the orthocarbonates, should also be accessible. A promising approach to synthesize, e.g., the alkali metal carbonates would be to apply high hydrostatic pressures during syntheses to the phase equilibria $M_2O + M_2(CO_3) \rightleftharpoons M_4(CO_4)$ (M = alkali metal). Since the hypothetical orthocarbonate would compete with high-pressure phases of the corresponding regular carbonates plus oxides, it is necessary to study the parts of the enthalpy surfaces of the M/C/O system with composition $M_4(CO_4)$, M_2O, and $M_2(CO_3)$ for many different pressures. In this way, one can theoretically establish the range of (thermodynamic) stability of the orthocarbonate phase versus decomposition into the corresponding oxide and carbonate as a function of applied pressure.

To achieve this, the enthalpy landscapes of M_2O [44], $M_2(CO_3)$ [177], and $M_4(CO_4)$ [24, 178] were investigated for many different pressures using simulated annealing and an empirical Coulomb-plus-Lennard–Jones potential.[14] All structure candidates in these systems were locally minimized on *ab initio* level in the Hartree–Fock approximation. Both for M_2O and $M_2(CO_3)$, the experimentally observed phases at standard and at high pressures were obtained, and in several of these 10 systems further high-pressure modifications were predicted to exist. Next, for each pressure, the thermodynamically stable modification was determined, and the enthalpy of $M_4(CO_4)$ was compared with the one of $M_2O + M_2(CO_3)$ as a function of pressure [178]. It was found that for all alkali metals, there should exist thermodynamically stable orthocarbonates at sufficiently high pressures, with the most easily accessible candidates being the potassium and rubidium orthocarbonates where the phase equilibrium is expected to switch to the orthocarbonate from the oxide-plus-carbonate in the range of 20–30 GPa.

4.4.1.5 Alkali Metal Sulfides M_2S (M = Li, Na, K, Rb, Cs)

Analogous predictions of the high-pressure phases were performed for the five alkali metal sulfides [43, 172]. Using empirical Coulomb-plus-Lennard–Jones potentials for the global search with simulated annealing followed by local optimizations on Hartree–Fock and density functional level, the experimentally observed modifications at standard pressure were found. Furthermore, several high-pressure phases, exhibiting the cotunnite and the Ni_2In-type were predicted. In parallel to high-pressure experiments, these predictions were verified for Li_2S [179], Na_2S [180], and K_2S [181]; the experimental confirmation for Rb_2S and Cs_2S has not yet been attempted.

---

14) In a first round of global optimizations, individual metal-, carbon-, and oxygen-atoms were used to describe atom configurations. After it turned out that the minimum configurations contained isolated trigonal CO_3- and tetrahedral CO_4-units, the latter at high pressures, a second round of global optimizations was performed, where CO_3- and CO_4-units were employed together with the metal atoms.

4.4.1.6 Boron Nitride

Another example of the global exploration of energy landscapes on *ab initio level*, i.e., without a first round of investigations using an empirical potential, is the application of simulated annealing to structure prediction for boron nitride [32], where several kinds of, mostly covalent, contributions to the total energy are present. The global searches were performed on both the Hartree–Fock and the density functional level,[15] using four formula units, starting from random atom arrangements in a large unit cell. After the low-lying minima had been identified, local optimizations using standard *ab initio* tolerance parameters followed, where both Hartree–Fock- and density functional methods were employed.

The BN system is particularly interesting as a test system, because the experimentally observed modifications include both layered structures (hexagonal BN) and three-dimensional networks (wurtzite- and sphalerite-type). In the global optimizations, all experimentally observed structure types were indeed found. In addition, several new modifications were predicted such as layered structures but with a stacking order different from the experimentally observed structure h-BN. Two other very interesting new low-energy modifications consist of three-dimensional BN-networks, showing the β-BeO structure, and the Al-partial structure in $SrAl_2$, respectively, with both BeO and the Al_2^{2-}-substructure of $SrAl_2$ being isoelectronic to BN.

4.4.1.7 Structure Prediction of SrO as Function of Temperature and Pressure

Predicting the possible (metastable) modifications of SrO as a function of temperature and pressure is essentially equivalent to determining the free enthalpy landscape(s) of SrO. This landscape was constructed by combining runs with simulated annealing, the ESA, and the threshold algorithm for a global exploration of the enthalpy landscapes [42, 182], where an empirical Coulomb-plus-Lennard–Jones potential served as an energy function. After a global optimization of the landscape using simulated annealing, the local minima identified during this stage were used as starting points for a large number of threshold runs at several different pressures [182]. This yielded an overview over both the low-lying local minima on the enthalpy landscapes and the barriers separating the different modifications. Next, the ESA was applied at standard pressure, and for a large number of different temperatures, in order to identify possible high-temperature phases. The potential energy and the radial distribution function served as indicator variables. All the structure candidates found with ESA turned out to be associated with individual local minima that had already been detected during simulated annealing or the threshold runs. Finally, the appearance of the melt phase was observed by checking the stability of

---

15) The searches with the Hartree–Fock approximation proved to be considerably faster (by a factor of about 4) than the ones using DFT-based energy calculations, because the former could better handle the task of converging the electronic structure calculations at low density for random structures. In general, one could speed-up the *ab initio* energy calculations by somewhat increasing the tolerance factors and thus greatly reducing the number of integrals required to calculate the total energy.

the underlying crystalline lattice of the rocksalt-type modification during very long MC-simulations for large simulation cells as a function of temperature.

In the fourth step, the free energies of the structure candidates found were computed in the quasi-harmonic approximation with the empirical potential, and also on the DFT-B3LYP level. Combining these free energies as a function of temperature with the energy barriers computed via the threshold algorithm

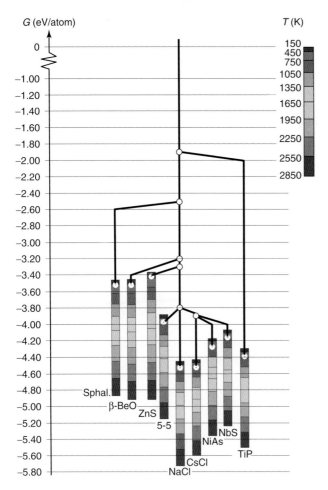

Figure 4.3 Free enthalpy landscape of SrO at $p = 0$ GPa for eight different temperatures ($T = 150$ K, ..., 2850 K) [42] computed using global landscape explorations followed by free energy calculations in the quasi-harmonic approximation on the empirical potential and *ab initio* level. The energetic contributions to the barriers stabilizing locally ergodic regions exhibiting different structure types are given by the energy difference between the minima (circles inside columns) and transition regions (white circles). Entropic barrier contributions (for a typical example see, e.g., [112]) are not shown to avoid overloading the figure. (Please find a color version of this figure on the color plates.)

resulted in the free enthalpy landscapes (Figure 4.3 depicts the case $p = 0$ GPa). These landscapes show that for the full range of temperatures and pressures investigated, the experimentally observed modifications, exhibiting the NaCl-type and the CsCl-type at standard and high pressure, respectively, corresponded to the global free energy minima on realistic observational time scales.

4.4.1.8 Phase Diagrams of the Quasi-Binary Mixed Alkali Halides

The computationally perhaps most expensive task is the prediction of a multinary phase diagram with no input except the identity of the participating atoms, suggesting that a modular approach should be used [7] : In the first step, the energy landscapes for many different compositions are explored using a robust empirical potential. In the second step, the many hundreds of structure candidates found during the global optimization are locally optimized on quantum mechanical level. A comprehensive structural analysis of the structure candidates must now follow, in order to decide whether we are dealing with solid solutions or ordered crystalline modifications.[16]

The crucial issue is whether so-called structure families exist among the minima observed for many different compositions which have essentially the same energy for a given composition [28, 48]. For solid solutions, this implies that the same overall cation–anion superstructure is present for many compositions, and the different types of cations and/or anions are randomly distributed over the cation or anion positions in the superstructure, respectively. If that is the case, the union of these local minima can be treated as a large locally ergodic region, and the free energy of this solid solution phase contains an entropy of mixing which favors the solution over ordered crystalline compounds which correspond to single minimum basins on the energy landscape. Finally, one can compute the Gibbs free energy via the convex hull method [185] using a combination of the ideal entropy of mixing and a Redlich–Kister polynomial Ansatz [186] for the enthalpy of mixing.

This approach has been applied to about 20 different quasi-binary alkali halide systems [28, 187–190], where the focus was on the low-temperature region of the phase diagram since the solidus–liquidus region was already known experimentally. The global landscape exploration for many different compositions was performed using simulated annealing and a Coulomb-plus-Lennard–Jones potential for the energy evaluation, followed by local optimizations using both Hartree–Fock and DFT methods. In all cases, the calculations correctly predicted whether a solid solution or ordered crystalline modifications were thermodynamically stable, and for those systems where the miscibility gap had been measured, the computed binodal was in good quantitative agreement with the experimental data [191] (the error in the critical temperatures of the computed miscibility gaps was estimated to be about ± 100 K), as is shown in Figure 4.4 for the system NaBr–LiBr [28, 192]. For those systems where crystalline modifications were predicted to be thermodynamically stable, these agreed with those already known from experiment, and several additional stable and metastable compounds could be predicted [187].

---

16) Such an analysis can be performed using e.g., the structure comparison algorithm CMPZ [183] implemented in the program KPLOT [184].

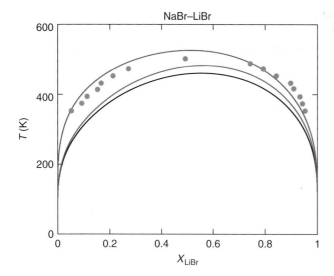

Figure 4.4 Low-temperature region of the phase diagram for the system NaBr–LiBr showing the miscibility gap in the system [28]. The gap was computed using global landscape explorations followed by the determination of free enthalpies employing both Hartree–Fock (black curve) and DFT–B3LYP (red curve) calculations. The blue curve is a fit to experimental data [191]; the yellow dots are experimental data points [192]. (Please find a color version of this figure on the color plates.)

4.4.2
Structure Prediction Employing Structural Restrictions

In this section, we show examples involving two types of structural restrictions on the allowed configuration space during structure predictions: primary or secondary building units, and underlying lattices on which (or near which) the atoms have to reside. Here, primary building units are groups of atoms, e.g., forming a complex ion or a whole molecule, that exist as individual groups of atoms in the final structure. In contrast, secondary building units (SBU) are groups of atoms, e.g., coordination polyhedra or cages in zeolites, that share atoms with other building units. We count these restricted searches as part of structure prediction, since these restrictions are made based on general chemical experience, in contrast to the prescription of the unit cell which implies that at least some powder diffraction data of the compound are available.

4.4.2.1 Complex Ions as Primary Building Units

As examples of structure prediction employing primary building units, we present two quasi-ionic systems, where the presence of strong covalent bonds among some of the atoms suggests that one should employ rigid building units instead of single atoms to describe the configurations during the global search. The price one pays is that one cannot a priori correctly assign the charge distribution of the building unit when computing the total energy of the test configurations during the random walk. Unless one can perform the calculation with *ab initio* energies or there are very strong arguments favoring a particular charge distribution, experience has shown that one should repeat the global searches for different charge distributions, in order to ensure that one does not overlook important structure candidates.

KNO$_2$ Here, a $(NO_2)^{1-}$ building unit was employed during simulated annealing runs [9, 33], with geometrical data taken from compounds listed in the ICSD. It was found that the most prominent structure candidates exhibit a distorted rock salt structure, if one considers only the centers of mass of the NO$_2$ groups and the potassium-atoms. All these minima taken together constitute a structure family, which at elevated temperatures forms the basis of a large locally ergodic region that corresponds to a high-temperature phase of the system. Such an "average" rock salt structure is also observed experimentally at room temperature for KNO$_2$, where one assumes either a positional or rotational disorder of the NO$_2$ groups to be present [193, 194]. Note that one will never observe such thermal disorder directly in the global optimizations, since the cost function always refers to $T = 0$. Thus, the structures one finds are always low-temperature structures, which in this case correspond to some low-symmetry modification of the highly symmetric high-temperature structure. To identify the high-temperature phase directly, one needs to perform long-time molecular dynamics or Monte Carlo simulations, as was done in Ref. [113], or employ the ESA. However, it is possible to study the activation barriers of the rotation of the NO$_2$-units around various axes of the unit, and one finds that two barriers of about 10 and 100 K, respectively, appear to dominate the dynamics. [48, 195] This suggests that at least one intermediary structure with limited rotational freedom of the NO$_2$-units should exist between the ordered global minimum and the freely rotating high-temperature structure. In the experiment, both a low-temperature structure (in space group $P2_1/c$) corresponding to one of the local minima seen on the energy landscape, and a structure with the NO$_2$-units rotating along the threefold axis of the crystal (space group $\bar{R3}$ m) at intermediary temperatures have been found. The theoretical results suggest that there might be a second, not yet observed, intermediary phase where the NO$_2$-units rotate along a twofold axis of the structure.

MgCN$_2$ Here, the carbon and nitrogen atoms were combined to a fixed $(CN_2)^{2-}$ unit. The geometry of the unit was taken as the average of structural elements found in various solid compounds containing CN$_2$ as a unit. The global optimizations with simulated annealing resulted in a large number of structure candidates [9], where two structures were most prominent: the experimentally observed structure [196], and a slightly more dense structure, similar to the Fe(S$_2$)-structure type, which might be thermodynamically stable at high pressures [24].

4.4.2.2 Molecular Crystals

An important class of crystalline compounds for which building units play a central role in predicting their structures are the molecular crystals [197–203], because the molecules in these crystals can essentially always be treated as indestructible units; one often even fixes the conformation of the molecules during the global search.

There are a number of critical issues that workers in this area have to contend with. (1) The number of local minima that have very similar energies is overwhelmingly large, and thus it is not easy to select the "relevant" structure candidates from among them. (2) Related to this problem is the difficulty to properly compute the total energy of the structures, both during the global searches and during the local optimization.[17] One typically uses simple empirical potentials to describe the intermolecular interactions during the global optimization stage.[18] (3) In principle, the molecules are flexible and not rigid, and treating them as inflexible during the global search can easily lead to important candidates being overlooked. (4) Due to the similarity in the ground-state energies of many structure candidates, the effect of the thermodynamic conditions (pressure, temperature) at which the crystal is synthesized can lead to a change in the ranking according to the free energies. (5) Finally, it is quite likely that with the system being able to "choose" from among many structures with nearly the same energies, the kinetic aspects of the synthesis process will end up controlling the structure of the molecular crystal found in the laboratory; especially, since the likelihood of formation of critical nuclei of the various phases is not necessarily correlated with the thermodynamic stability of the corresponding macroscopic crystal.

Seen from the perspective of extended solids, it comes as a surprise that a small set of space groups (18) suffices to describe over 90% of all molecular crystals found so far [205]. Furthermore, over 90% of these crystals contain only one molecule in the asymmetric unit [206]. Since this observation is often exploited when solving molecular crystal structures, it is natural that theoretical structure determination and prediction have proceeded by taking this (statistical) symmetry information into account when generating structure candidates [207]. As a consequence, in most studies, the random-walker-based global minimizations tend to be combined with massive exhaustive searches, and the use of simulated annealing and other stochastic-walker-based algorithms has not been very critical to the success of the prediction. Thus, we present only two examples; for further cases, we refer to one of the many useful review articles that have appeared in recent years [9, 10, 199, 200, 208–210], and to the results of the relatively recent third blind test [202].

---

17) Due to the fact that high-quality ab-initio calculations of crystals containing both inter- and intramolecular interactions are still rather challenging, one has usually employed refined empirical potentials for the refinement optimizations. Such potentials are either fitted to experiment or to results from ab-initio calculations on molecules.

18) Such an approach can be quite successful. For example, for molecular nitrogen a simulated annealing study, using simple Coulomb and Lennard–Jones potentials and a quadrupolar charge distribution within the N_2-building unit succeeded in finding the experimentally observed low-temperature $Pa3$ phase of solid nitrogen [9, 33, 204].

Conformation-family Monte Carlo study Pillardy et al. [88] employed the so-called conformation-family Monte Carlo algorithm to predict the crystal structure of nine small organic molecules such as benzene, pyrimidine, dimethoxymethane, or formamidoxime, using several empirical potentials (Coulomb-, Lennard–Jones-, and Buckingham-type potentials, plus torsional energy terms) to compute the energy during the global search, and several refined potentials (AMBER [211], W99 [212]) during the local optimizations. Both the location, orientation, and conformation of the molecule, and the shape and size of the unit cell were modified during the global search. No symmetry constraints were applied. For the rigid molecules, their shape was taken from crystal structures where these molecules were present if such structures were available.

In the four studies involving rigid molecules, the energy of the lowest (presumably global) minimum was lower than the energy of the experimental structure (computed with the same refined potential), indicating that the global optimization technique was working quite well although the energy functions employed were not yet accurate enough. In nearly all cases, i.e., combinations of potentials employed, the experimental structure was found as a local minimum, but often not as the global minimum. When searching for crystal structures for the five flexible molecules, the success rate was not so high: in most instances, the experimental structure was not found during the global search, and the energy of the lowest minimum found was higher than the one belonging to the experimental structure in several cases.

It is interesting to note that in the majority of cases, the global search found local minima whose energies were lower than those of the experimentally observed crystal structures; sometimes the experimental modifications were not even among the top 10 minima by energy. This appears to be a general problem for molecular crystal structure prediction, in contrast to the prediction of extended crystalline solids, where the experimental structure is usually among the four or five lowest minima in energy (if found at all), and often corresponds to the global minimum on *ab initio* level, the exact rank depending on the type of *ab initio* method employed for the local optimization.[19]

Phenobarbital The general strategy pursued by Day et al. [213] in their investigation of the energy landscape of phenobarbital, a medium-sized molecule with two flexible torsion angles, provides a good example of the currently popular approach to structure prediction of molecular crystals: In a first step, the molecule is analyzed on refined potential and/or *ab initio* (here DFT) level, and possible conformations of the molecule that are expected to be most relevant are determined. Then, a divide-and-conquer approach is used: global optimizations using simulated annealing are performed for several (here up to nine) of the most common space

---

19) However, one should not become overconfident: it is quite possible that once one starts dealing with, e.g., multi-atom-type intermetallic compounds, a similar plethora of minima of nearly equal energy are going to be found – one only needs to think of the many local minimum configurations associated with the locally ergodic region of a solid solution!

groups in molecular crystals, and a few (here up to three) molecules in the asymmetric unit – for each of several different conformations (here five, with three corresponding to local minima and two to saddle points on the energy hypersurface of the molecule).[20] The molecular conformation is kept fixed, but the position and orientation of the molecule in the cell, and the cell parameters are allowed to vary during the run. The minima found are now locally optimized using refined potentials, e.g., W99 [212] or CVFF-950 [214], and polynomial interpolants of the DFT-based hypersurface of the molecule, where both the positions and the conformations of the molecules were allowed to change.

Similar to the experience in the preceding study, it turned out that e.g., the experimentally observed structure "III" ranked only 15th by energy among the local minima found for $Z' = 1$, indicating that even the highly refined potentials fitted to *ab initio* data or combinations of *ab initio* and potential energies were not yet sufficiently accurate. Neither of the two experimental structures "I" and "II" of phenobarbital was found during the global optimization. They exhibit both $Z' = 3$, and their energies proved to be considerably lower than the energies of the lowest minima found during the global search. It is encouraging that the search produced many energetically competitive structures for $Z' = 1, 2, 3$, which can serve as possible candidates for a fourth known modification of phenobarbital whose structure has not yet been determined. Nevertheless, even this rather extensive careful study (over 620 000 crystal structures were locally minimized) encountered serious problems when dealing with a flexible molecule with nonnegligible hydrogen bonding.

4.4.2.3 Zeolites
An important class of compounds for which structures have been predicted are the zeolites and zeolite-analogs [10, 22, 24, 209, 215–218]. However, when using unrestricted global optimization techniques such as simulated annealing, zeolite framework prediction encounters serious problems. The reason is that we cannot expect the simulated system to produce the zeolite framework "on its own" within a reasonable time when starting from individual atoms, although structure determination of zeolites with restricted energy landscapes has already been successfully performed [219]. Addressing this issue has led to the development of the AASBU-procedure [22], and, alternatively, the use of restrictions on the overall cell volume of the allowed configurations.

AASBU-procedure Draznieks *et al.* [22] proposed the "automated assembly of secondary building blocks" (AASBU), which proceeds by taking structural elements like coordination polyhedra and joining them at their corners, edges and faces, in order to generate new structures. The interactions among the SBUs during the generation phase are based on a simple Lennard–Jones-like force field, while potentially "merging" atoms are provided with a strong attractive potential term, in order to encourage the linking of the polyhedra. The centers of the polyhedra

---

20) In the phenobarbital study, the asymmetric unit used contained more than one molecule ($Z' > 1$) only for the two most common space groups, $P2_1/c$ and $P\bar{1}$.

interact via a strong repulsive potential preventing an overlap of the SBUs beyond the merging distance. As starting configurations, a random arrangement of a fixed number of rigid SBUs is chosen. Periodic boundary conditions are employed, and commonly a space group is prescribed. simulated annealing is used for the global optimization, where the SBUs are allowed to rotate. Furthermore, the cell parameters can be changed, and the distances among the SBUs adjusted.

In order to generate zeolite-like structures, ML_4-tetrahedra were picked as SBUs in this study, in agreement with experimental data, where the interactions were chosen to favor corner-sharing networks. One or two SBU per asymmetric unit, and a selected set of space groups were employed. The authors found the expected structure types, e.g., the GME, FAU, RHO, and LTL frameworks. For LTL-like frameworks, two new candidate structures were found, and an energy minimization with GULP [220] using default potentials showed that their lattice energies were only slightly higher than that of LTL itself.

These results and subsequent successes [209] are quite encouraging. An unknown quantity is the effect of fixing the space group during the optimization – it is not clear, whether the above frameworks could be generated in $P1$, too.

Exclusion zones and density restrictions An alternative approach to zeolite prediction consists in varying the positions of individual atoms and the cell parameters while ensuring that the overall density of the configuration remains within a given interval during the simulated annealing [48]. After local optimization (quench) on empirical potential level, the resulting structures (for a SiO_2-zeolite) were quite reasonable (see Figure 4.5), underlinn the large degree of freedom the system has in forming porous structures.

Finally, another alternative method consists in introducing stationary or mobile exclusion zones together with individual atoms or larger building units. Such work has been mostly performed with genetic algorithms by Woodley et al. [216, 221–223], although simulated annealing has also been employed, e.g., for the study of possible zeolite-type structures in the SiO_2-analog BeF_2 using spherical mobile zones of varying diameter [224]. In the course of these latter studies, it was observed that there is often a tendency for mobile exclusion zones to cluster during the global exploration stage. This can sometimes lead to structures with isolated columnar pores or, in the extreme case, to alternating sheets of essentially bulk material (e.g., some SiO_2-type modification, or other structures consisting of corner- or edge-connected BeF_4-tetrahedra) and layers of exclusion zones, i.e., empty space.[21]

4.4.2.4 Phase Diagrams Restricted to Prescribed Sublattices

A special case of restricted structure prediction is the prediction of phase diagrams where all atoms have to be located on the sites of a predefined set of sublattices

---

21) One should note that such slabs of bulk material always constitute relatively low-lying local minima that are energetically competitive with three-dimensional porous structures when enforcing a low overall density for the system, especially for large numbers of atoms/simulation cells.

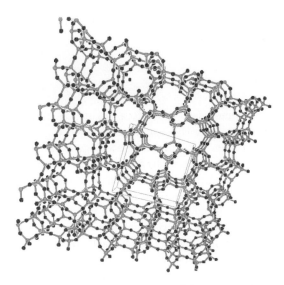

Figure 4.5 Zeolite-like structure candidate generated using freely movable silicon (light spheres) and oxygen (dark spheres) atoms and free cell parameters, under the condition that the overall density of the structure must stay within a certain range typical for zeolites [48].

usually known from experiment [225–233]. Taking this as a starting point, a number of studies have tried to predict phase diagrams including both line compounds and solid solutions, on a restricted energy landscape where all atoms have to be located on the sites of the predefined lattice or set of sublattices, and the moveclass consists of atoms exchanging positions, possibly followed by a local minimization after each exchange [23].

MgO–CaO, MgO–MnO, and (Ca,Mg)CO$_3$ Examples employing Monte Carlo simulations can be found, e.g., in work by Allan and coworkers [23, 234–236], who have performed semigrand canonical Monte Carlo simulations using ionic interatomic model potentials to compute phase diagrams of MgO–CaO, MgO–MnO, and (Ca,Mg)CO$_3$ [23]. In the experiment, all these systems exhibit solid-solution phases. No ordered phases were observed, and the agreement with experiment was within about 100 K for the miscibility gap.

M$_x$CoO$_2$ (M = Li, Na) Similarly, Ceder and coworkers investigated the temperature-concentration (T, x) phase diagrams of Li$_x$CoO$_2$ [237] and Na$_x$CoO$_2$ [238], where the energy calculations were performed using effective cluster interaction terms and DFT calculations. MC simulations were used to explore the possible alkali metal positions on the appropriate sublattice in the alkali metal cobalt oxide structure. For certain compositions, ordered crystalline phases were observed, and furthermore several phase transitions as a function of temperature and concentration were predicted. Overall, the computed phase diagrams were in good agreement with experiment.

4.4.3
Structure Determination

While the true structure prediction described above is a fascinating area of fundamental research in crystallography, chemistry and materials science, a more restricted somewhat modified application of the same methodology looks ready to become an invaluable tool in applied crystallography and solid-state chemistry: structure determination from limited experimental information using energy landscapes.

The high pace of modern research has led to a multitude of microcrystalline solids that have been synthesized, or generated e.g., via high-pressure experiments [239, 240], but for which no single crystal X-ray data are available. Instead, one solves their structures based on powder diffraction data, where usually the most difficult step is the generation of structure candidates for the final refinement.

From the point of view of applied mathematics, finding such candidates can be treated as an optimization problem, with cost functions resembling, at least to a certain degree, the energy landscapes of solids. Thus, simulated annealing and other structure prediction techniques should be a natural tool to use for structure determination, too.

4.4.3.1 Structure Determination using Experimental Cell Information
Much of the earliest work in the general area of structure determination has incorporated information about the size and shape of the unit cell, perhaps even including symmetry information from systematic extinction of reflexes, and the knowledge of the content of the unit cell [14, 15, 241, 242]. In particular, global optimizations using simulated annealing on a bond-valence-plus-Coulomb-potential based cost-function landscape were successfully performed by Pannetier et al. [14] for the determination of the atom position of NbF_4 and K_2NiF_4. Similarly, the polymorphs of TiO_2 were determined by Freeman et al. [15] for given unit cell size, shape, and content, where the authors proceeded in two steps: First, the unit cell was kept fixed and purely repulsive interaction among the atoms were used as cost function during the global search with simulated annealing. This was followed by local minimizations using a realistic empirical potential.

4.4.3.2 Reverse Monte Carlo Method and Pareto Optimization
The preceding methodology introduces only part of the experimental information available from e.g., powder diffraction data into the global search on the energy landscape. Alternatively, one can primarily optimize the agreement between experimental measurements and the structure using the so-called Reverse Monte Carlo (RMC) method, where the cost function is the difference between the measured powder diffractogram and the one computed for a simulated atomic configuration. Reference to the energy of the structure is only made obliquely via penalty terms

in the cost function that keep atoms from overlapping. This approach has been employed quite often to generate models of amorphous compounds [243, 244], usually employing simulated annealing as the global optimization technique. An implementation of this procedure for crystalline substances is the program ESPOIR [245], which has been successfully tested for a number of ionic, quasi-ionic and molecular crystals.[22]

A more balanced approach uses a good energy function as in general structure prediction, but modifies the energy landscape through explicit incorporation of experimental data by adding the difference R_B (sometimes denoted the R-value) between the measured Bragg intensities and the ones calculated for the current atomic configuration, to the potential energy. An early implementation of this idea appeared in work on the structure determination of zeolites [247, 248]. Here, Deem and Newsam [247] defined a cost function (called "figure-of-merit") as the sum of energy terms and the difference between the measured and computed powder diffractogram for the atoms in the centers of the tetrahedral building units (called "T-atoms") in a zeolite, ignoring the connecting oxygen atoms. The interaction energy between the T-atoms consisted of statistical potential terms that reproduced the distance and angle distributions among neighboring T-atoms. In addition to the knowledge of the unit cell and its T-atom content, space group and some general zeolite framework information was included in the optimization process, which employed simulated annealing. Using the distance constraints and powder data, the observed zeolite framework was found among the many candidates generated, in about 90% of the examples.

Several years later [249], this approach was generalized to a full Pareto optimization [250], where the diffractogram and the empirical potential energy enter the cost function on an equal footing. The structure is optimized both with respect to the energy and the diffractogram,

$$E = \lambda E_{\text{pot}} + (1-\lambda) R_B, (0 \leq \lambda \leq 1) \tag{4.9}$$

This approach has been tested successfully for a large number of ionic, quasi-ionic and metallic systems [249, 251], where simulated annealing was used as the global optimization tool.[23] Simple two-body potentials with Coulomb- and Lennard–Jones-terms served as energy functions; such simple potentials were sufficient because the combination of experimental input and theoretical energy function delivered a high synergy by eliminating many unrealistic local minima on the energy landscape. One up-to-date implementation, the program ENDEAVOUR [253], has already been very successful in "real-life" applications, generating convincing structure candidates for such different systems as K_2CN_2 [254], sulfur

---

22) Another implementation of the RMC method has been used for lattice and magnetic structure analysis, where RMC is combined with neutron powder diffraction data [246].

23) Such Pareto optimizations to solve synthesized structures from powder data were also performed using genetic algorithms [252].

[255], Na$_3$PSO$_3$ [256], Ag$_2$NiO$_2$ [257], Ag$_2$PdO$_2$ [258], GaAsO$_4$ [259], ammonium metatungstate [260], the zeolite-like structure Na$_{1-x}$Ge$_{3+z}$ [261], Tl$_2$CS$_3$ [262], and BiB$_3$O$_6$ [263].

4.5
Evaluation and Outlook

4.5.1
State-of-the-Art

"Beware the claims of the producer!" [264] is a word of caution one should heed quite generally when reviewing the performance of global optimization techniques, such as the many variants of the simulated annealing algorithm we have presented in the preceding sections. As we indicated in Section 4.3, between the moveclass of the random walker, the temperature schedule, the size of the ensemble of walkers, and the degree to which landscape information is incorporated, there exists a large amount of freedom to tune and optimize any particular algorithm under consideration. The general strength of simulated annealing and its relatives shows itself most clearly if one tries to find the deepest minima on an energy landscape about which barely any information is available besides the energy function itself. In its most basic version, this very generic search algorithm is very easy to implement, and since its invention, it has been employed for many combinatorial and other multiminima optimization problems as "a first attempt."[24]

Considering the general task of finding most of the relevant deep-lying local minima of a crystalline solid, or quite generally a chemical system, using simulated annealing and related methods, what is the current state of the field? Of course, the size of the system that can be handled depends on the computer power available; thus, let us consider a researcher who has access to a small cluster of PCs such that several hundred or perhaps several thousand (e.g., if many different compositions or pressures are to be studied) single simulated annealing runs are feasible within a reasonable amount of time. Under these conditions, for binary and ternary systems, standard simulated annealing can deal with about 20–30 atoms/simulation cell when using empirical potentials, and about 8–12 atoms/cell using *ab initio* energy calculations. More refined, but often computationally more

---

24) Such an unbiased search often does not appear to be very efficient, though much more appealing to the purist. (Clearly, if the only goal is to identify the deepest minimum, one should use any method at hand to accomplish this task!) For a more measured technical, in some ways perhaps more scientific approach to judging the quality of an algorithm, one wants to be able to measure the (computational) work during the exploration campaign in a way that takes into account not only the classical floating point operations or function evaluations (here energy and gradient calculations), but also moveclass modifications, memory requirements and incorporation of landscape information (gathered in earlier studies or as part of the current run). Only in this fashion one would be able to fairly judge the amount of this "generalized" effort involved for a given outcome of the search, or in the dual problem, properly evaluate the quality of the results for a fixed amount of generalized effort.

expensive and/or algorithmically more complex, variants such as optimal cycling or basin hopping are on average more successful than standard simulated annealing for up to 40–50 atoms/cell (for empirical potentials), but beyond this number, none of the methods appears to be truly reliable. This applies even more strongly if one attempts to deal with large quaternary or even more diverse systems, whose landscapes are dominated by minima corresponding to essentially amorphous configurations: unless the energy function possesses structure-directing features, e.g., the energy hypersurface contains large funnels guiding the search toward the most important local minima, one has to ask oneself "Am I feeling lucky today?" when exploring such systems.

Regarding the use of building units, it is somewhat difficult to fairly evaluate the progress in the field of molecular crystal structure prediction since a large amount of the search effort tends to be devoted to systematically scanning the energy landscape, and the use of divide-and-conquer approaches by prescribing the allowed symmetry of the system. Thus the relevance of, e.g., choosing simulated annealing for part of the global optimization is difficult to assess. A similar caveat applies to zeolite structure prediction where often symmetry constraints are employed during the simulated annealing.

We have not discussed the many applications of simulated annealing to clusters and polymers. They are in a class by themselves because these systems do not obey periodic boundary conditions and their landscapes are often both more diverse, with many more local minima and homotops, and at the same time more controllable, as far as the application and refinement of optimization algorithms are concerned, than crystalline systems. But it appears that a well-tuned simulated annealing algorithm, using, e.g., an optimal cycling temperature schedule, can hold its own with more refined methods such as basin hopping or genetic algorithms also for clusters [69]. This observation most likely holds quite generally: one can probably expect that random-walker-based algorithms whose moveclass and temperature schedules are well-optimized (?!), and similarly optimized genetic algorithms, taboo searches or other general search algorithms will yield comparable results.

4.5.2
Future

"Optimize the optimization algorithm!" will surely remain one of the watchwords of the future of structure prediction, irrespective of whether we are dealing with simulated annealing-type algorithms or not. A corollary of this is that newer and bigger computers are not the magic bullet, since the difficulty of solving typical structure prediction problems grows exponentially with system size. However, we can glean some hope from the fact that, to paraphrase Einstein, "nature might be subtle but not malicious": there often appear to be approximately hierarchical elements in those periodic structures that contain many atoms/primitive unit cell, and this must surely be reflected in the properties of the corresponding energy landscape. Once we can identify such general features of the landscape of crystalline solids, including information about e.g., the growth law of the local densities of states that can influence the progress of the walker [53, 54, 112, 265], we should be able to greatly improve the performance of various simulated annealing-type algorithms by e.g., employing adaptive moveclasses.

Hierarchical or divide-and-conquer approaches that work by restricting the allowed configuration space during the simulated annealing often appear to be able to deal with large simulation cells or molecules. However, one always runs a considerable risk of overlooking fascinating structure candidates, for example, when basing one's decision to employ certain building units only on, e.g., database information about related chemical systems. This is especially true when one predominantly relies on chemical intuition instead of mathematical information about the shape of the energy landscape. On the other hand, once one has established that all the low-lying minima of a system found for small simulation cells correspond to structures that exhibit certain invariant structural elements, one can employ these with a reasonably good conscience during a simulated annealing run.

Finally, minima are not everything that one cares about when studying an energy landscape, even if one only wants to determine structure candidates. As we pointed out earlier, identifying complex locally ergodic regions, determining the local free energy of the various hypothetical modifications and their kinetic stability requires landscape information beyond the local minima, such as (generalized) barriers and local densities of states. This will presumably become even more important in the future when one tries to address the issue of how to deal efficiently with systems that exhibit controlled disorder or possess large locally ergodic regions at elevated temperatures. As indicated earlier, there are already many algorithms available for this purpose, but most are still rather clumsy and inefficient. Optimizing these exploration tools will clearly be a major enterprise in the future.

References

1. Corey, E. J. (1967) *Pure Appl. Chem.*, **14**, 19.
2. Corey, E. J. (1991) *Angew. Chem. Int. Ed. Eng.*, **30**, 455.
3. Maddox, J. (1988) *Nature*, **335**, 201.
4. Cohen, M. L. (1989) *Nature*, **338**, 291.
5. Hawthorne, F. C. (1990) *Nature*, **345**, 297.
6. Catlow, C. R. A. and Price, G. D. (1990) *Nature*, **347**, 243.
7. Schön, J. C. and Jansen, M. (1996) *Angew. Chem. Int. Ed. Eng.*, **35**, 1286.
8. Jansen, M. (2002) *Angew. Chem. Int. Ed.*, **41**, 3747.
9. Schön, J. C. and Jansen, M. (2001) *Z. Krist.*, **216**, 307.
10. Woodley, S. M. and Catlow, C. R. A. (2008) *Nature Mater.*, **7**, 937.
11. Schön, J. C. and Jansen, M. (2009) *Int. J. Mat. Res.*, **100**, 135.
12. Jansen, M., (2008) in *Turning Points in Solid-State, Materials and Surface Science*, (eds K. M. Harris and P. Edwards) RSC Publishing, Cambridge, UK, p. 22.
13. Liu, A. Y. and Cohen, M. L. (1990) *Phys. Rev. B*, **41**, 10727.
14. Pannetier, J., Bassas-Alsina, J., Rodriguez-Carvajal, J., and Caignaert, V. (1990) *Nature*, **346**, 343.
15. Freeman, C. M., Newsam, J. M., Levine, S. M., and Catlow, C. R. A. (1993) *J. Mater. Chem.*, **3**, 531.
16. Schön, J. C. and Jansen, M. (1994) *Ber. Bunsenges.*, **98**, 1541.

17. Boisen, M. B., Gibbs, G. V., and Bukowinski, M. S. T. (1994) *Phys. Chem. Miner.*, **21**, 269.
18. Schön, J. C. and Jansen, M. (1995) *Comp. Mater. Sci.*, **4**, 43.
19. Bush, T. S., Catlow, C. R. A., and Battle, P. D. (1995) *J. Mater. Chem.*, **5**, 1269.
20. Putz, H., Schön, J. C., and Jansen, M. (1998) *Comp. Mater. Sci.*, **11**, 309.
21. Woodley, S. M., Battle, P. D., Gale, J. D., and Catlow, C. R. A. (1999) *Phys. Chem. Chem. Phys.*, **1**, 2535.
22. Mellot-Draznieks, C., Newsam, J. M., Gorman, A. M., Freeman, C. M., and Ferey, G. (2000) *Angew. Chem. Int. Ed. Eng.*, **39**, 2270.
23. Allan, N. L., Barrera, G. D., Lavrentiev, M. Y., Todorov, I. T., and Purton, J. A. (2001) *J. Mater. Chem.*, **11**, 63.
24. Mellot-Draznieks, C., Girard, S., Ferey, G., Schön, J. C., Čančarević, Ž., and Jansen, M. (2002) *Chem. Eur. J.*, **8**, 4102.
25. Winkler, B., Pickard, C. J., Milman, V., and Thimm, G. (2001) *Chem. Phys. Lett.*, **337**, 36.
26. Goedecker, S. (2004) *J. Chem. Phys.*, **120**, 9911.
27. Oganov, A. R. and Glass, C. W. (2006) *J. Chem. Phys.*, **124**, 244704.
28. Schön, J. C., Pentin, I. V., and Jansen, M. (2006) *Phys. Chem. Chem. Phys.*, **8**, 1778.
29. Doll, K., Schön, J. C., and Jansen, M. (2007) *Phys. Chem. Chem. Phys.*, **9**, 6128.
30. Woodley, S. M. (2004a) *Struct. Bonding*, **110**, 95.
31. Schön, J. C., Doll, K., and Jansen, M. (2010) *phys stat. sol.*, (b), **247**, 23.
32. Doll, K., Schön, J. C., and Jansen, M. (2008) *Phys. Rev. B*, **78**, 144110.
33. Schön, J. C. and Jansen, M. (1999) *Acta Cryst A (Suppl.)*, **55**.
34. Le Bail, A. and Calvayrac, F. (2006) *J. Solid State Chem.*, **179**, 3159.
35. Kirkpatrick, S., Gelatt, C. D. Jr., and Vecchi, M. P. (1983) *Science*, **220**, 671.
36. Czerny, V. (1985) *J. Optim. Theo. Appl.*, **45**, 41.
37. Holland, J. H. (1975) *Adaptation in Natural and Artificial Systems* University of Michigan Press, Ann Arbor.
38. Sibani, P., Schön, J. C., Salamon, P., and Andersson, J.-O. (1993) *Europhys. Lett.*, **22**, 479.
39. Schön, J. C. (1996) *Ber. Bunsenges.*, **100**, 1388.
40. Schön, J. C., Putz, H., and Jansen, M. (1996) *J. Phys.: Cond. Matt.*, **8**, 143.
41. Laio, A. and Parrinello, M. (2002) *Proc. Nat. Acad. Sci.*, **99**, 12562.
42. Schön, J. C., Čančarević, Ž. P., Hannemann, A., and Jansen, M. (2008) *J. Chem. Phys.*, **128**, 194712.
43. Schön, J. C., Čančarević, Ž., and Jansen, M. (2004) *J. Chem. Phys.*, **121**, 2289.
44. Čančarević, Ž., Schön, J. C., and Jansen, M. (2006a) *Phys. Rev. B*, **73**, 224114.
45. Goldstein, M. (1969) *J. Chem. Phys.*, **51**, 3728.
46. Stillinger, F. and Weber, T. A. (1982) *Phys. Rev. A*, **25**, 978.
47. Wales, D. J. (2003) *Energy Landscapes with Applications to Clusters, Biomolecules and Glasses* Cambridge University Press Cambridge.
48. Schön, J. C. and Jansen, M., in (2005) *Mat. Res. Soc. Symp. Proc. Vol. 848: Solid State Chemistry of Inorganic Materials V*, (eds J. Li, N. E. Brese, M. G. Kanatzidis, and M. Jansen) MRS, Warrendale, pp. 333–344.
49. Schön, J. C., Wevers, M. A. C., and Jansen, M. (2003) *J. Phys.: Cond. Matter*, **15**, 5479.
50. Schön, J. C., Wevers, M., and Jansen, M. (2001a) *J. Phys. A: Math. Gen.*, **34**, 4041.
51. Hoffmann, K. and Schön, J. C. (2005) *Found. Phys. Lett.*, **18**, 171.
52. Geman, S. and Geman, D. (1984) *IEEE T. Pattern Anal*, **6**, 721.
53. Schön, J. C. (1997) *J. Phys. A: Math. Gen.*, **30**, 2367.
54. Salamon, P., Sibani, P., and Frost, R. (2002) *Facts, Conjectures, and Improvements for Simulated Annealing* SIAM Monographs, Philadelphia.
55. Pickard, C. J. and Needs, R. J. (2006) *Phys. Rev. Lett.*, **97**, 045504.
56. Kroll, P. and Hoffmann, R. (1998) *Angew. Chem. Int. Ed. Eng.*, **37**, 2527.

57. Strong, R. T., Pickard, C. J., Milman, V., Thimm, G., and Winkler, B. (2004) *Phys. Rev. B*, **70**, 045101.
58. Le Bail, A. (2005) *J. Appl. Cryst.*, **38**, 389.
59. Curtarolo, S., Morgan, D., Persson, K., Rodgers, J., and Ceder, G. (2003) *Phys. Rev. Lett.*, **91**, 135503.
60. Delsante, G. G. S., Borzone, G., Asta, M., and Ferro, R. (2006) *Acta Mater.*, **54**, 4977.
61. Li, Z. and Scheraga, H. A. (1987) *Proc. Nat. Acad. Sci.*, **84**, 6611.
62. Buch, V., Martonak, R., and Parrinello, M. (2006) *J. Chem. Phys.*, **124**, 204705.
63. Putz, H., Schön, J. C., and Jansen, M. (1995) *Ber. Bunsenges.*, **99**, 1148.
64. Huitema, H. and Eerden, J.P. v.d. (1999) *J. Chem. Phys.*, **110**, 3267.
65. Wales, D. J. and Doye, J. P. K. (1997) *J. Phys. Chem.*, **101**, 5111.
66. Iwamatsu, M. and Okabe, Y. (2004) *Chem. Phys. Lett.*, **399**, 396.
67. Delamarre, D. and Virot, B. (1998) *RAIRO – Rech. Oper. Oper. Res.*, **32**, 43.
68. Möbius, A., Neklioudov, A., Diaz-Sanchez, A., Hoffmann, K. H., Fachat, A., and Schreiber, M. (1997) *Phys. Rev. Lett.*, **79**, 4297.
69. Möbius, A., Hoffmann, K., and Schön, J. C., in (2004) Complexity, Metastability and Nonextensivity, (eds C. Beck, G. Benedek, A. Rapisarda, and C. Tsallis), *International School of Solid State Physics*, World Scientific, Singapore, pp. 215–219.
70. Ruppeiner, G., Pedersen, J. M., and Salamon, P. (1991) *J. Phys. I*, **1**, 455.
71. Salamon, P., Nulton, J., Robinson, J., Pedersen, J. M., Ruppeiner, G., and Liao, L. (1988) *Comp. Phys. Comm.*, **49**, 423.
72. Hoffmann, K. H., Würtz, D., de Groot, C., and Hanf, M., in (1991) *Parallel and Distributed Optimization*, (eds M. Grauer and D. B. Pressmar) Springer, Heidelberg, p. 154.
73. Metropolis, N., Rosenbluth, A. W., Rosenbluth, M. N., Teller, A. H., and Teller, E. (1953) *J. Chem. Phys.*, **21**, 1087.
74. Szu, H. and Hartley, R. (1987) *Phys. Lett. A*, **122**, 157.
75. Tsallis, C. (1988) *J. Stat. Phys.*, **52**, 479.
76. Tsallis, C. and Stariolo, S. A. (1996) *Physica A*, **233**, 395.
77. Dueck, G. and Scheuer, T. (1990) *J. Comp. Phys.*, **90**, 161.
78. Glover, F. (1990) *Interfaces*, **20**, 74.
79. Cvijovic, D. and Klinowski, J. (1995) *Science*, **267**, 664.
80. Ji, M. and Klinowski, J. (2006) *Proc. Roy. Soc. A*, **462**, 3613.
81. Huber, T., Torda, A., and van Gunsteren, W. F. (1994) *J. Comput. Aided Mol. Des.*, **8**, 695.
82. Chandy, J. A., Kim, S., Ramkumar, B., Parkes, S., and Banerjee, P. (1997) *IEEE Trans. Comp. Aided Des. ICS*, **16**, 398.
83. Zimmermann, T. and Salamon, P. (1992) *Int. J. Comp. Math.*, **42**, 21.
84. Ma, J., Hsu, D., and Straub, J. E. (1993) *J. Chem. Phys.*, **99**, 4024.
85. Roitberg, A. and Elber, R. (1991) *J. Chem. Phys.*, **95**, 9277.
86. Straub, J. E. and Karplus, M. (1990) *J. Chem. Phys.*, **94**, 6737.
87. Kim, J. G., Fukunishi, Y., Kidera, A., and Nakamura, H. (2004) *Chem. Phys. Lett.*, **392**, 34.
88. Pillardy, J., Arnautova, Y. A., Czaplewski, C., Gibson, K. D., and Scheraga, H. A. (2001) *Proc. Nat. Acad. Sci.*, **98**, 12351.
89. Venkatnathan, A. and Voth, G. A. (2005) *J. Chem. Theo. Comp.*, **1**, 36.
90. Huber, T. and van Gunsteren, W. F. (1998) *J. Phys. Chem.*, **102**, 5937.
91. Berg, B. A., Nogushi, H., and Okamoto, Y. (2003) *Phys. Rev. E*, **68**, 036126.
92. Itoh, S. G. and Okamoto, Y. (2004) *Chem. Phys. Lett.*, **400**, 308.
93. Kim, J. and Keyes, T. (2004) *J. Chem. Phys.*, **121**, 4237.
94. Berg, B. A. and Neuhaus, T. (1992) *Phys. Rev. Lett.*, **68**, 9.
95. Frantz, D. D., Freeman, D. L., and Doll, J. D. (1990) *J. Chem. Phys.*, **93**, 2769.
96. Voter, A. F. (1997a) *J. Chem. Phys.*, **106**, 4665.
97. Voter, A. F. (1997b) *Phys. Rev. Lett.*, **78**, 3908.

98. Wawak, R. J., Pillardy, J., Liwo, A., Gibson, K. D., and Scheraga, H. A. (1998) *J. Phys. Chem. A*, **102**, 2904.
99. Zhang, Y., Kihara, D., and Skolnick, J. (2002) *Proteins: Struct. Funct. Gen.*, **48**, 192.
100. Zhu, Z., Tuckerman, M. E., Samuelson, S. O., and Martyana, G. J. (2002) *Phys. Rev. Lett.*, **88**, 100201.
101. Merlitz, H. and Wenzel, W. (2002) *Chem. Phys. Lett.*, **362**, 271.
102. Hamelberg, D., Morgan, J., and McCommon, J. A. (2004) *J. Chem. Phys.*, **120**, 11919.
103. Hamelberg, D., Shen, T., and McCommon, J. A. (2005) *J. Chem. Phys.*, **122**, 241103.
104. Zhang, W. and Duan, Y. (2006) *Protein: Eng. Design Struct.*, **19**, 55.
105. Wenzel, W. and Hamacher, K. (1999) *Phys. Rev. Lett.*, **82**, 3003.
106. Hamacher, K. (2006) *Europhys. Lett.*, **74**, 944.
107. Hamacher, K. (2007) *Physica A*, **378**, 307.
108. Cheng, L., Cai, W., and Shao, X. (2005) *ChemPhysChem*, **6**, 261.
109. Dueck, G. (1993) *J. Comp. Phys.*, **104**, 86.
110. Sibani, P. and Schriver, P. (1994) *Phys. Rev. B*, **49**, 6667.
111. Sibani, P., Pas, R. v. d., and Schön, J. C. (1999) *Comp. Phys. Comm.*, **116**, 17.
112. Wevers, M. A. C., Schön, J. C., and Jansen, M. (1999) *J. Phys.: Cond. Matt.*, **11**, 6487.
113. Duan, C. G., Mei, W. N., Smith, R. W., Liu, J., Ossowski, M. M., and Hardy, J. R. (2001) *Phys. Rev. B*, **63**, 144105.
114. Asker, C., Belonoshko, A. B., Mikhaylushkin, A. S., and Abrikosov, I. A. (2008) *Phys. Rev. B*, **77**, 220102.
115. Chodera, J. D., Singhal, N., Pande, V. S., Dill, K. A., and Swope, W. C. (2007) *J. Chem. Phys.*, **126**, 155101.
116. Berry, R. S. (1993) *Chem. Rev.*, **93**, 2379.
117. Banerjee, A., Adams, N., Simmons, J., and Shepard, R. (1985) *J. Phys. Chem.*, **89**, 52.
118. Berry, R. S., Davis, H. L., and Beck, T. L. (1988) *Chem. Phys. Lett.*, **147**, 13.
119. Fischer, S. and Karplus, M. (1992) *Chem. Phys. Lett.*, **194**, 252.
120. Ionova, I. V. and Carter, E. A. (1993) *J. Chem. Phys.*, **98**, 6377.
121. Wales, D. J. (1993) *J. Chem. Soc. Farad. Trans.*, **89**, 1305.
122. Mauro, J. C., Loucks, R. J., and Balakrishnan, J. (2005) *J. Phys. Chem. A*, **109**, 9578.
123. Mills, G. and Jonsson, H. (1994) *Phys. Rev. Lett.*, **72**, 1124.
124. Tanaka, H. (2000) *J. Chem. Phys.*, **113**, 11202.
125. Weinan, E., Weiqing, R., and Vanden-Eijnden, E. (2002) *Phys. Rev. B*, **66**, 052301.
126. Barkema, G. T. and Mousseau, N. (1996) *Phys. Rev. Lett.*, **77**, 4358.
127. Wei, G., Mousseau, N., and Derreumaux, P. (2002) *J. Chem. Phys.*, **117**, 11379.
128. Mauro, J. C., Loucks, R. J., and Balakrishnan, J. (2006) *J. Phys. Chem. B*, **110**, 5005.
129. Schlegel, H. B. (2003) *J. Comp. Chem.*, **24**, 1514.
130. Santiso, E. E. and Gubbins, K. E. (2004) *Mol. Sim.*, **30**, 699.
131. Zahn, D. and Leoni, S. (2004) *Phys. Rev. Lett.*, **92**, 250201.
132. Carter, E. A., Ciccotti, C., Hynes, J. T., and Kapral, R. (1989) *Chem. Phys. Lett.*, **156**, 472.
133. Faradijan, A. K. and Elber, R. (2004) *J. Chem. Phys.*, **120**, 10880.
134. Grubmüller, H. (1995) *Phys. Rev. E*, **52**, 2893.
135. Bolhuis, P. G., Dellago, C., and Chandler, D. (1998) *Faraday Discuss.*, **110**, 421.
136. Dellago, C., Bolhuis, P., Csajka, F. S., and Chandler, D. (1998) *J. Chem. Phys.*, **108**, 1964.
137. Dellago, C., Bolhuis, P., and Geissler, P. L., (2006) in *Computer Simulations in Condensed Matter: From Materials to Chemical Biology*, (eds M. Ferrario, G. Ciccotti, and K. Binder), Springer, New York, p. 124.
138. Pratt, L. R. (1986) *J. Chem. Phys.*, **85**, 5045.
139. Peters, B., Liang, W. Z., Bell, A. T., and Chakraborty, A. (2003) *J. Chem. Phys.*, **118**, 9533.

140. Chen, L. Y. and Nash, P. L. (2003) *J. Chem. Phys.*, **119**, 12749.
141. Dimelow, R. J., Bryce, R. A., Masters, A. J., Hillier, I. H., and Burton, N. A. (2006) *J. Chem. Phys.*, **124**, 114113.
142. Wales, D. J., Doye, J. P. K., Miller, M. A., Mortenson, P. N., and Walsh, T. R., (2000) in *Advances in Chemical Physics, volume 115*, (eds I. Prigogine and S. A. Rice) Wiley, New York, pp. 1–111.
143. Ferrando, R., Jellinek, J., and Johnston, R. L. (2008) *Chem. Rev.*, **108**, 845.
144. Johnston, R. L. and Roberts, C., (2003) in *Soft computing approaches in chemistry*, (eds H. M. Cartwright and L. M. Sztendera) Springer, Berlin, pp. 161–204.
145. Hartke, B. (2004) *Structure and Bonding*, **110**, 33.
146. Car, R., Parrinello, M., and Andreoni, W., in (1987) *Proceedings of the 1st NEC Symposium on Fundamental Approaches to New Material Phases*, (eds S. Sugano, Y. Nishina, and S. Ohnishi), Tokyo, pp. 134–141.
147. Hohl, D., Jones, R., Car, R., and Parrinello, M. (1987) *Chem. Phys. Lett.*, **139**, 540.
148. Jones, R. O. and Seifert, G. (1992) *J. Chem. Phys.*, **96**, 7564.
149. Phillips, N. G., Conover, C. W. S., and Bloomfield, L. A. (1991) *J. Chem. Phys.*, **94**, 4980.
150. Lai, S. K., Hsu, P. J., Wu, K. L., Liu, W. K., and Iwamatsu, M. (2002) *J. Chem. Phys.*, **117**, 10715.
151. Goedecker, S., Hellmann, W., and Lenosky, T. (2005) *Phys. Rev. Lett.*, **95**, 055501.
152. Gehrke, R. and Reuter, K. (2009) *Phys. Rev. B*, **79**, 085412.
153. Dill, K. A., Bromberg, S., Yue, K., Fiebig, K. M., Yee, D. P., Thomas, P. D., and Chan, H. S. (1995) *Prot. Science*, **4**, 561.
154. Koningsveld, R. and Stockmeyer, W. H. (2001) *Polymer Phase Diagrams*, Oxford University Press, Oxford
155. Muthukumar, M. (2005) *Adv. Polym. Sci.*, **191**, 241.
156. Rao, F. and Caflisch, A. (2004) *J. Mol. Biol.*, **342**, 299.
157. Moult, J. (2005) *Curr. Opin. Struct. Biol.*, **15**, 285.
158. Petrey, D. and Honig, B. (2005) *Molecular Cell*, **20**, 811.
159. Floudas, C. A., Fung, H. K., McAllister, S. R., Mönnigmann, M., and Rajgaria, R. (2006) *Chem. Eng. Sci.*, **61**, 966.
160. Adcock, S. A. and McCammon, J. A. (2006) *Chem. Rev.*, **106**, 1589.
161. Prentiss, M. C., Zong, C., Hardin, C., Eastwood, M. P., and Wolynes, P. G. (2006) *J. Chem. Theo. Comp.*, **2**, 705.
162. Zhang, Y. (2008) *Curr. Opin. Struct. Biol.*, **18**, 342.
163. Haas, H. and Jansen, M. (1999) *Angew. Chem. Int. Ed. Eng.*, **38**, 1910.
164. Claeyssens, F., Freeman, C. L., Allan, N. L., Sun, Y., Ashfold, M. N. R., and Harding, J. H. (2005) *J. Mater. Chem.*, **15**, 139.
165. Čančarević, Ž., Schön, J. C., and Jansen, M. (2008) *Chem. Asian. J.*, **3**, 561.
166. Liebold-Ribeiro, Y., Fischer, D., and Jansen, M. (2008) *Angew. Chem. Int. Ed.*, **47**, 4428.
167. Bach, A., Fischer, D., and Jansen, M. (2009) *Z. Anorg. Allg. Chem.*, **635**, 2406.
168. Fischer, D. and Jansen, M. (2002a) *J. Am. Chem. Soc.*, **124**, 3488.
169. Brese, N. E. and O'Keeffe, M., in (1992) *Struct. Bonding*, Springer, Heidelberg, p. 307.
170. Jansen, M. and Schön, J. C. (1998) *Z. Anorg. Allg. Chem.*, **624**, 533.
171. Schön, J. C., Wevers, M. A. C., and Jansen, M. (2000) *Solid State Sci.*, **2**, 449.
172. Schön, J. C., Wevers, M. A. C., and Jansen, M. (2001b) *J. Mater. Chem.*, **11**, 69.
173. Fischer, D. and Jansen, M. (2002b) *Angew. Chem. Int. Ed.*, **41**, 1755.
174. Vajenine, G. V., Wang, X., Efthimiopoulus, I., Karmakar, S., Syassen, K., and Hanfland, M. (2008) *Z. Anorg. Allg. Chem.*, **634**, 2015.
175. Vajenine, G. V., Wang, X., Efthimiopoulus, I., Karmakar, S., Syassen, K., and Hanfland, M. (2009) *Phys. Rev. B*, **79**, 224107.

176. Ozolins, V., Majzoub, E. H., and Wolverton, C. (2008) *Phys. Rev. Lett.*, **100**, 135501.
177. Čančarević, Ž., Schön, J. C., and Jansen, M. (2006b) *Z. Anorg. Allg. Chem.*, **632**, 1437.
178. Čančarević, Ž., Schön, J. C., and Jansen, M. (2007) *Chemistry Europ. J.*, **13**, 7330.
179. Grzechnik, A., Vegas, A., Syassen, K., Loa, I., Hanfland, M., and Jansen, M. (2000) *J. Solid State Chem.*, **154**, 603.
180. Vegas, A., Grzechnik, A., Syassen, K., Loa, I., Hanfland, M., and Jansen, M. (2001) *Acta Cryst. B*, **57**, 151.
181. Vegas, A., Grzechnik, A., Hanfland, M., Mühle, C., and Jansen, M. (2002) *Solid State Sci.*, **4**, 1077.
182. Schön, J. C. (2004) *Z. Anorg. Allg. Chem.*, **630**, 2354.
183. Hundt, R., Schön, J. C., and Jansen, M. (2006) *J. Appl. Cryst.*, **39**, 6.
184. Hundt, R., (1979) KPLOT: A Program for Plotting and Investigation of Crystal Structures, University of Bonn, Germany, Version 9, 2007.
185. Voronin, G. (2003) *Russian J. Phys. Chem.*, **77**, 1874.
186. Redlich, O. and Kister, A. T. (1948) *Ind. Eng. Chem.*, **40**, 345.
187. Pentin, I. V., Schön, J. C., and Jansen, M. (2007) *J. Chem. Phys.*, **126**, 124508.
188. Schön, J. C., Pentin, I. V., and Jansen, M. (2007) *J. Phys. Chem. B*, **111**, 3943.
189. Schön, J. C., Pentin, I. V., and Jansen, M. (2008) *Solid State Sci.*, **10**, 455.
190. Pentin, I. V., Schön, J. C., and Jansen, M. (2008) *Solid State Sci.*, **10**, 804.
191. Sangster, J. and Pelton, A. (1987) *J. Phys. Chem. Ref. Data*, **16**, 509.
192. Doornhof, D., Wijk, H. V., and Hoonk, H. (1984) *Thermochim. Acta*, **76**, 171.
193. Solbakk, J. K. and Stromme, K. O. (1969) *Acta Chem. Scand.*, **23**, 300.
194. Onoda-Yamamuro, N., Honda, H., Ikeda, R., Yamamuro, O., Mtsuo, T., Oikawa, K., Kamiyama, T., and Izumi, F. (1998) *J. Phys. Cond. Matter*, **10**, 3341.
195. Schön, J. C., Salamon, P., and Jansen, M., (2010) in prep.
196. Berger, U. and Schnick, W. (1994) *J. Alloys Comp.*, **206**, 179.
197. Gavezzotti, A. (1991) *J. Amer. Chem. Soc.*, **113**, 4622.
198. Holden, J. R., Du, Z., and Ammon, H. L. (1993) *J. Comp. Chem.*, **14**, 422.
199. Gdanitz, R. J. (1997) in *Theoretical Aspects and Computer Modeling*, (eds A. Gavezzotti) Wiley, New York, pp. 185.–201.
200. Verwer, P. and Leusen, F. J. J. (1998) in *Reviews of Computational Chemistry*, (eds K. B. Lipkowitz and D. B. Boyd), Wiley-VCH, New York, vol. 12, pp. 327–365.
201. van Eijck, B. P. and Kroon, J. (1999) *J. Comp. Chem.*, **20**, 799.
202. Day, G. M. et al., (2005) *Acta Cryst. B*, **61**, 511.
203. Karamertzanis, P. G. and Pantelides, C. C. (2007) *Mol. Phys.*, **105**, 273.
204. Schön, J. C. (1999) Vortr. Chem. Doz. Tag., Oldenburg (ADUC) p. B15.
205. Baur, W. H. and Kassner, D. (1992) *Acta Cryst. B*, **48**, 356.
206. Padmaja, N., Ramakumar, S., and Wiswamitra, M. A. (1990) *Acta Cryst. A*, **46**, 725.
207. Filippini, G. and Gavezzotti, A. (1992) *Mol. Cryst. Liq. Cryst.*, **219**, 37.
208. Price, S. L. (2004) *Adv. Drug Deliv. Rev.*, **56**, 301.
209. Mellot-Draznieks, C. (2007) *J. Mater. Chem.*, **17**, 4348.
210. Price, S. L. (2008) *Phys. Chem. Chem. Phys.*, **10**, 1996.
211. Weiner, S., Kollman, P. A., Nguyen, D. T., and Case, D. A. (1986) *J. Comp. Chem.*, **7**, 230.
212. Williams, D. E. (2001) *J. Comp. Chem.*, **22**, 1154.
213. Day, G. M., Motherwell, W. D. S., and Jones, W. (2007) *Phys. Chem. Chem. Phys.*, **9**, 1693.
214. Dauber-Osguthorphe, P., Roberts, V. A., Osguthorpe, D. J., Wolff, J., Genest, M., and Hagler, A. T. (1988) *Proteins: Struct. Funct. Genet.*, **4**, 31.
215. Treacy, M. M. J., Randall, K. H., Rao, S., Perry, J. A., and Chadi, D. J. (1997) *Z. Krist.*, **212**, 768.
216. Woodley, S. M., Catlow, C. R. A., Battle, P. D., and Gale, J. D. (2004a) *Chem. Comm.*, **2004**, 22.
217. Engel, N. (1991) *Acta Cryst. B*, **47**, 849.

218. Zwijnenburg, M. A., Bromley, S. T., Foster, M. D., Bell, R. G., Delgado-Friedrichs, O., Jansen, J. C., and Maschmeyer, T. (2004) *Chem. Mater.*, **16**, 3809.
219. Deem, M. W. and Newsam, J. M. (1989) *Nature*, **342**, 260.
220. Gale, J. D. (1997) *J. Chem. Soc. Farad. Trans.*, **93**, 629.
221. Woodley, S. M., Battle, P. D., Gale, J. D., and Catlow, C. R. A. (2004b) *Phys. Chem. Chem. Phys.*, **6**, 1815.
222. Woodley, S. M. (2004b) *Phys. Chem. Chem. Phys.*, **6**, 1823.
223. Woodley, S. M. (2007) *Phys. Chem. Chem. Phys.*, **9**, 1070.
224. Schön, J. C. and Jansen, M., unpublished.
225. Curtarolo, S., Kolmogorov, A. N., and Cocks, F. H. (2005) *Comp. Coupl. Phase Diagr. Thermochem.*, **29**, 155.
226. Sanati, M., Wang, L. G., and Zunger, A. (2003) *Phys. Rev. Lett.*, **90**, 045502.
227. Laradji, M., Landau, D. P., and Dünweg, B. (1995) *Phys. Rev. B*, **51**, 4894.
228. Hirschl, R., Hafner, J., and Jeanvoine, Y. (2001) *J. Phys.: Cond. Matter*, **13**, 3545.
229. Wolverton, C., Ozolins, V., and Asta, M. (2004) *Phys. Rev. B*, **69**, 144109.
230. Fuks, D., Dorfman, S., Piskunov, S., and Kotomin, E. A. (2005) *Phys. Rev. B*, **71**, 014111.
231. Allan, N. L., Barrera, G. D., Lavrentiev, M. Y., Freeman, C. L., Tordov, I. T., and Purton, J. A. (2006) *Comp. Mater. Sci.*, **36**, 42.
232. Purton, J. A., Allan, N. L., Lavrentiev, M. Y., Todorov, I. T., and Freeman, C. L. (2006) *Chem. Geol.*, **225**, 176.
233. Bärthlein, S., Hart, G. L. W., Zunger, A., and Müller, S. (2007) *J. Phys.: Cond. Matter*, **19**, 032201.
234. Lavrentiev, M. Y., Allan, N. L., Barrera, G. D., and Purton, J. A. (2001) *J. Phys. Chem. B*, **105**, 3594.
235. Marquez, F. M., Cienfuegos, C., Pongsal, B. K., Lavrentiev, M. Y., Allan, N. L., Purton, J. A., and Barrera, G. D. (2003) *Model. Simul. Mater. Sci. Eng.*, **11**, 115.
236. Lavrentiev, M. Y., Allan, N. L., and Purton, J. A. (2003) *Phys. Chem. Chem. Phys.*, **5**, 2190.
237. Van der Ven, A., Aydinol, M. K., Ceder, G., Kresse, G., and Hafner, J. (1998) *Phys. Rev. B*, **58**, 2975.
238. Hinuma, Y., Meng, Y. S., and Ceder, G. (2008) *Phys. Rev. B*, **77**, 224111.
239. Liu, L. and Bassett, W. A. (1986) *Elements, Oxides and Silicates. High-Pressure Phases with Implications for the Earth's Interior* Oxford University Press, New York.
240. R. J. Hemley, ed. (1998) *Ultrahigh-Pressure Mineralogy: Physics and Chemistry of the Earth's Deep Interior*, vol. 37 of Reviews in Mineralogy. The Mineralogical Society of America, Washington, DC.
241. Catlow, C. R. A., Bell, R. G., and Gale, J. D. (1994) *J. Mater. Chem.*, **4**, 781.
242. Belashchenko, D. K. (1994) *Inorg. Mater. (Engl. Trans.)*, **30**, 966.
243. Kaplow, R., Rowe, T. A., and Averbach, B. L. (1968) *Phys. Rev.*, **168**, 1068.
244. McGreevy, R. L. (1997) in *Computer Modelling in Inorganic Crystallography*, (eds C. R. A. Catlow) Academic Press, San Diego, pp. 151–184.
245. LeBail, A. (2000) in *Proc. EPDIC-7* preprint.
246. Mellergard, A. and McGreevy, R. L. (1999) *Acta Cryst. A*, **55**, 783.
247. Deem, M. W. and Newsam, J. M. (1992) *J. Am. Chem. Soc.*, **114**, 7189.
248. Falcioni, M. and Deem, M. W. (1999) *J. Chem. Phys.*, **110**, 1754.
249. Putz, H., Schön, J. C., and Jansen, M. (1999) *J. Appl. Cryst.*, **32**, 864.
250. Pareto, V. (1896/97) *Cours D'Economie Politique*.
251. Coelho, A. A. (2000) *J. Appl. Cryst.*, **33**, 899.
252. Lanning, O. J., Habershon, S., Harris, K. D. M., Johnston, R. L., Kariuki, B. M., Tedesco, E., and Turner, G. W. (2000) *Chem. Phys. Lett.*, **317**, 296.
253. Putz, H. (2000) Endeavour 1.0, Crystal Impact GbR, Bonn.
254. Becker, M. and Jansen, M. (2000) *Solid State Sci.*, **2**, 711.
255. Crichton, W. A., Vaughan, G. B. M., and Mezouar, M. (2001) *Z. Krist.*, **216**, 417.

256. Pompetzki, M. and Jansen, M. (2002) *Z. Anorg. Allg. Chem.*, **628**, 641.
257. Schreyer, M. and Jansen, M. (2002) *Angew. Chem. Int. Ed.*, **41**, 643.
258. Schreyer, M. and Jansen, M. (2001) *Solid State Sci.*, **3**, 25.
259. Santamaria-Perez, D., Haines, J., Amador, U., Moran, E., and Vegas, A. (2006) *Acta Cryst. B*, **62**, 1019.
260. Christian, J. B. and Whittingham, M. S. (2008) *J. Solid State Chem.*, **181**, 1782.
261. Beekman, M., Kaduk, J. A., Huang, Q., Wong-Ng, W., Yang, Z., Wang, D., and Nolas, G. S. (2007) *Chem. Comm.*, **2007**, 837.
262. Beck, J. and Benz, S. (2009) *Z. Anorg. Allg. Chem.*, **635**, DOI:10.1002/zaac.200801408
263. Dinnebier, R. E., Hinrichsen, B., Lennie, A., and Jansen, M. (2009) *Acta Cryst. B*, **65**, 1.
264. Salamon, P. (1988), priv. comm.
265. Schön, J. C. and Sibani, P. (1998) *J. Phys. A: Math. Gen.*, **31**, 8165.

5
Simulation of Structural Phase Transitions in Crystals: The Metadynamics Approach
Roman Martoňák

5.1
Introduction

Computational materials science nowadays plays an important role in technology and applications. Besides studying properties of existing materials, there is also lot of interest in search for new materials. It is well known that high pressure can drive in crystals structural transitions to denser structures which may have substantially different properties. Often, the barriers that have to be crossed in structural transformations are large which results in the widespread existence of metastable phases. This phenomenon can actually be useful since it could enable a high-pressure structure to be quenchable to low pressure which offers a practical way to prepare materials having unusual properties under normal conditions.

The problem of studying crystalline structures under variable pressure and temperature is clearly twofold. One part consists in determination of the phase diagram and is in principle purely thermodynamical. This has traditionally been known as the crystal structure prediction (CSP) problem – for a given pressure P and temperature T, one has to determine what is the stable crystalline structure of the given substance. Since the problem can be formulated as an optimization problem, namely search for the global minimum of the Gibbs free energy, or, in the simpler case at $T = 0$, of enthalpy, it has been a testbed for the whole spectrum of optimization algorithms, such as simulated annealing [1], simple random search [2], evolutionary algorithms [3–5], etc. Recently a remarkable progress has been achieved due to the evolutionary search algorithm by Oganov and Glass [4], in particular at $T = 0$. Once the candidate structures are known, free energy calculation techniques allow us, at least in principle, to calculate the phase boundaries and therefore determine the thermodynamical phase diagram.

While the first part of the problem focuses on the question "What structure should be created under given conditions?", the other part of the problem addresses instead the question "How it is actually created?" This is relevant for several reasons. Perhaps most importantly, the structure created in the experiment might not be the one that is thermodynamically stable under given conditions, but rather a metastable one, because of kinetic reasons. Therefore, by concentrating

Modern Methods of Crystal Structure Prediction. Edited by Artem R. Oganov
Copyright © 2011 WILEY-VCH Verlag GmbH & Co. KGaA, Weinheim
ISBN: 978-3-527-40939-6

on the process itself, rather than merely on its thermodynamically predicted outcome, it should also be possible to provide a more accurate answer to the question "What structure is actually created?" Besides that, understanding of the microscopic mechanisms of the transformation on the atomistic level is likely to be useful for experimental preparation of the new material, in order to drive the system toward the desired transition. Last but not least, the understanding of the microscopic mechanisms of structural transformations is intrinsically interesting, since extracting this information from experiment remains very difficult. In this respect simulations represent a particularly valuable complement of experiment.

Clearly, elucidating the processes of structural transformations in their full complexity is at present a too ambitious goal. Under real experimental conditions one can expect that all kinds of defects of the ideal crystalline structure will play a role in the process of nucleation. Besides that, as we will discuss later, the ubiquitous timescale gap problem manifests here in a dramatic way and has to be addressed.

In this chapter we discuss the approach to solving the second part of the problem, based on the metadynamics algorithm. The paper is organized as follows. In the second section we discuss the general aspects of the problem of simulating structural transitions in crystals. In the third section we present the metadynamics-based algorithm for simulation of structural transformations and in the fourth section we discuss its practical application. In the fifth section we illustrate the results obtained with the algorithm for several selected systems. Finally, in the last section we draw some conclusions and discuss the outlook.

5.2
Simulation of Structural Transformations

In the process of structural transformation the unit cell of the crystalline structure typically undergoes a substantial change and any method aiming at simulation of the process must allow for such modification. Clearly, in a constant volume molecular dynamics (MD) or Monte Carlo (MC) simulation where the supercell is chosen to be commensurate with the unit cell of the initial structure it is unlikely that it would be commensurate also with the final structure. This mismatch will result in a large free energy penalty for the final structure that will most likely prevent the transition from taking place. The first technique that addressed this problem was the constant-pressure MD by Parrinello and Rahman [6, 7] which introduced the variable cell. The supercell was treated as a dynamical variable allowed to fluctuate and to adapt according to the arrangement of atoms. This setup allowed naturally for simulation of structural transitions in crystals under conditions of hydrostatic as well as nonhydrostatic stress. It has been successfully applied to many systems described by classical force fields. Later, in combination with *ab initio* calculations [8] the technique allowed first-principles MD simulations of pressure-induced structural transitions and thus gained a much better predictive power.

As mentioned above, however, this kind of simulations suffers severely from the timescale gap problem. Most structural transitions are of first order (as a consequence of the lack of group–subgroup relation between the symmetry groups) and therefore proceed by nucleation and growth. In this process a free energy barrier which separates the initial from the final structure must be crossed. If the barrier is large in comparison to the thermal energy $k_B T$, the spontaneous crossing of the barrier is a rare event and occurs on the time scale which is very long compared to typical time scales of atomic vibrations. This occurs also in experiment, in particular in case of the reconstructive phase transitions where strong covalent bonds are broken across the transition, and some transitions would not even occur on the experimental time scale unless the kinetics is accelerated by increasing the temperature. In simulation, of course, the situation is much worse, since the time scale over which we can observe the evolution of the system is many orders of magnitude shorter than in experiment.

It is well known that simulation of such activated processes poses a problem if the barrier height is substantially larger than the thermal energy $k_B T$. In fact, in simulation of structural transitions we encounter precisely this situation. While in the real system nucleation is likely to start at structural defects (or at surface), where the symmetry is already broken locally, in simulation one typically uses a small system (at most few hundred atoms in the case of *ab initio* simulations) arranged as supercell with periodic boundary conditions. Under these conditions where neither surface nor defects are present it is very unlikely that a spontaneous nucleation event will occur and structural transformations occur typically in a collective manner, all atoms undergoing the same displacement at the same time. The barrier associated with such a collective mechanism could be substantially higher than the one corresponding to proper nucleation, which tends to make the timescale gap problem even worse. The most simple solution here is to simply increase the pressure beyond the thermodynamic transition pressure. This would decrease the barrier and increase the probability that it will be spontaneously crossed in the simulation time. When the pressure is high enough and the barrier becomes close to zero, the system gets close to the point of mechanical instability and the transition occurs on the fast time scale, typically in few picoseconds. This time scale can be nowadays easily reached even in *ab initio* simulations of systems consisting of several hundreds of atoms. The problem is, however, that the pressure at which the system loses mechanical stability could be much larger (often by an order of magnitude) than the equilibrium transition pressure. The transition observed under the conditions of strong overpressurization is likely to be different from the one observed at equilibrium, and might even result in the creation of a different structure. This clearly has a negative impact on the predictive power of the method, and even if the correct structure is created the mechanism might be substantially different from the one that would operate at equilibrium conditions. It is clear that any method aimed at improving this difficulty has to address the timescale gap problem and introduce some mechanism accelerating the barrier crossing.

We would like to stress that we do not aim here at simulating the structural transitions in crystals in full complexity, including nucleation processes. Instead we are looking for a practical technique allowing the simulation of the idealized model represented by a relatively small supercell with periodic boundary conditions, yet being able to observe the transitions that include barrier crossing without the need for an excessive overpressurization.

5.3
The Metadynamics-Based Algorithm

The metadynamics algorithm invented by Laio and Parrinello [9] represents a generic approach to the simulation of activated processes, such as chemical reactions, first-order phase transitions, or protein folding. It belongs to the family of methods using biasing potential which is added to the original free energy with the aim to lower the barrier and thus accelerate its crossing. This class of techniques includes, e.g., local elevation [10], conformational flooding [11], or hyperdynamics [12]. The difference among these techniques consists in the type of the space in which the biasing potential is applied as well as in the manner in which it is constructed (the details are found in the original papers). In metadynamics the biasing potential is constructed in the space of a suitably chosen order parameter with low dimensionality. This important step reduces the dimensionality of the search space in a substantial manner and makes the problem more easily manageable. The idea of working in the space of order parameter is based on the assumption of time separation. One assumes that it is possible to split the microscopic degrees of freedom of the problem into two groups – the fast ones which can be equilibrated on the simulation time scale, and the slow ones which require excessive time. In the ideal case the order parameter – which can consist of several components – should include all slow degrees of freedom. The order parameter is then treated by metadynamics while the remaining fast degrees of freedom are treated by a standard sampling method such as MD or MC.

While it is possible to invent various kinds of order parameters which could drive the structural transitions between different crystalline phases, it is also possible to use a simple and generic one which turns out to work well in many cases. The idea is based on the separation of time scales between acoustic and optical phonons. While optical phonons essentially represent internal vibrations of atoms inside the unit cell, acoustic phonons are rather related to the fluctuations of the unit cell of the crystalline structure. Since the latter degrees of freedom are slower, it is possible to construct an order parameter based on the unit cell components and use it in metadynamics. This approach is in spirit close to that of the Parrinello–Rahman (PR) method, where the variable unit cell is able to distinguish between different crystalline structures. In the following we show how this idea can be implemented in a practical scheme. We start with the original version proposed in Ref. [13] and afterward present the improved version from Ref. [14].

In the PR method [6] one applies periodic boundary conditions defined by the simulation box edges $\vec{a}, \vec{b}, \vec{c}$ and uses the 3×3 matrix $\mathbf{h} = (\vec{a}, \vec{b}, \vec{c})$ as collective coordinate, or order parameter. The matrix $\mathbf{h}$ is treated as a dynamical variable coupled to the atomistic degrees of freedom under the condition of constant average pressure. Since the matrix $\mathbf{h}$ contains besides 6 degrees of freedom related to the shape of the cell (3 cell lengths amd 3 angles) also 3 global rotations, it is useful to exclude these 3 rotational degrees of freedom from the problem. In Ref. [15] it was proposed to restrict the matrix $\mathbf{h}$ to the upper triangular form which effectively freezes the box rotation.

In such a case we can define our six-dimensional order parameter as a vector $\tilde{\mathbf{h}} = (h_{11}, h_{22}, h_{33}, h_{12}, h_{13}, h_{23})^T$. Following the generic metadynamics algorithm [9] we define a discrete dynamics by the equations

$$\tilde{\mathbf{h}}^{t+1} = \tilde{\mathbf{h}}^t + \delta h \frac{\phi^t}{|\phi^t|} \tag{5.1}$$

Here, the driving force $\phi^t = -\frac{\partial G^t}{\partial \tilde{\mathbf{h}}}$ is derived from a modified Gibbs potential $\mathcal{G}^t$ which includes a history-dependent term

$$\mathcal{G}^t(\tilde{\mathbf{h}}) = \mathcal{G}(\tilde{\mathbf{h}}) + \sum_{t' < t} W e^{-\frac{|\tilde{\mathbf{h}} - \tilde{\mathbf{h}}^{t'}|^2}{2\delta h^2}} \tag{5.2}$$

The history-dependent term [10] in $\mathcal{G}^t(\tilde{\mathbf{h}})$ pushes the system out of the local minimum. The first derivative of the free energy with respect to the order parameter can be expressed as

$$\frac{\partial \mathcal{G}}{\partial h_{ij}} = V \left[\mathbf{h}^{-1}(\sigma + P) \right]_{ji} \tag{5.3}$$

and requires only an evaluation of the stress tensor σ by means of an NVT ensemble average over a relatively short MD (or Monte Carlo) simulation.

The above version of the algorithm has been successfully applied to several systems [16–20]. In some cases, however, the use of the matrix $\mathbf{h}$ directly as the order parameter may not be convenient since the shape of the free energy well in the $\mathbf{h}$ coordinates may be quite anisotropic. Besides crystal anisotropy resulting in different stiffness with respect to the different components of $\mathbf{h}$, there is also a strong coupling among the components of $\mathbf{h}$. In particular, the energy cost of a deformation where the system is compressed along all axes (volume compression) is typically much higher than the cost of a deformation where compression along one axis is compensated by expansion in the perpendicular directions (volume-conserving deformation). The basin of attraction of a crystal structure is therefore likely to be narrow in the direction of volume change and long in the perpendicular directions. Proper exploration of such landscape would require the Gaussian size to be small with respect to the narrow direction. Consequently, filling the well would require a large number of metasteps, which is not practical. If, on the other hand, the Gaussian size is not small enough, the metadynamics algorithm does not guarantee that the system escapes from the basin of attraction of the initial crystal structure via the lowest energy path. In the extreme case of a large Gaussian (we note that the Gaussian size is equal to the metadynamics step), the system may escape in a random direction. This is likely to be important in systems with complex free energy surfaces where there are many competing pathways leading out of the initial free energy basin. Some of these pathways may lead to crystalline structures while others may result, e.g., in amorphization; therefore, it is important to perform the exploration of the free energy surface in a proper way so that the pathways crossing the low barriers are followed.

The shape of the bottom of the well is described by its curvature. Close to a given equilibrium crystal structure characterized by a matrix $\tilde{\mathbf{h}}^0$ the Gibbs free energy can be expanded to second order:

$$\mathcal{G}(\tilde{\mathbf{h}}) \approx \mathcal{G}(\tilde{\mathbf{h}}0) + \frac{1}{2}(\tilde{\mathbf{h}} - \tilde{\mathbf{h}}0)^T \mathbf{A}(\tilde{\mathbf{h}} - \tilde{\mathbf{h}}0) \tag{5.4}$$

The Hessian matrix

$$A_{ij} = \partial^2 \mathcal{G}(\tilde{\mathbf{h}})/\partial \tilde{h}_i \partial \tilde{h}_j \Big|_{\tilde{\mathbf{h}}_0} \tag{5.5}$$

can be calculated from the finite differences of the stress tensor at different values of $\mathbf{h}$, making use of Eq. (5.3). At equilibrium the matrix $\mathbf{A}$ has positive real eigenvalues $\{\lambda^i\}$ and can be diagonalized by an orthogonal matrix $\mathbf{O}$.

For illustration we show here an example from *ab initio* calculation for silicon. Table 5.1 shows the eigenvalues and eigenvectors of the Hessian matrix (5.5) corresponding to a supercell consisting of 64 atoms in the cubic diamond structure at a temperature $T = 0$ K and pressure $p = 100$ kbar. We note the large spread of eigenvalues as well as the fact that the eigenvector corresponding to the largest eigenvalue points along the direction (111 000). Since for an upper triangular matrix the volume is simply expressed as $V = \det \mathbf{h} = \tilde{h}_1 \tilde{h}_2 \tilde{h}_3$, the direction of the volume gradient in the $\tilde{\mathbf{h}}$ space is $\nabla V = (\tilde{h}_2 \tilde{h}_3, \tilde{h}_1 \tilde{h}_3, \tilde{h}_1 \tilde{h}_2, 0, 0, 0)$, which is approximately parallel to the direction (111 000) if the diagonal elements of $\mathbf{h}$ are not too different.

Table 5.1 Eigenvalues (in units of (kbar Å)) and corresponding eigenvectors of the Hessian matrix (5.5) for a sample of silicon consisting of 64 atoms in the cubic diamond structure at temperature $T = 0$ K and pressure $p = 100$ kbar.

Eigenvalues					
6439.8	6487.9	11858.0	11914.0	12110.7	62638.7

Eigenvectors					
0.7915	−0.2048	−0.0050	0.0048	−0.0015	0.5758
−0.5717	−0.5812	−0.0012	0.0058	0.0012	0.5791
−0.2161	0.7875	0.0069	−0.0074	−0.0032	0.5771
0.0012	0.0025	0.6297	0.7506	−0.2003	−0.0021
−0.0027	−0.0001	−0.3348	0.0296	−0.9418	−0.0018
0.0044	−0.0125	0.7009	−0.6601	−0.2699	0.0004

In our experience this is typical also for noncubic supercells; the largest eigenvalue is mainly associated with volume change and is usually larger than the smallest one by an order of magnitude and more.

In order to treat all degrees of freedom on equal footing it is convenient to diagonalize the quadratic form (5.4). This can be achieved by expressing the variables $\tilde{\mathbf{h}}$ in terms of new collective variables $\mathbf{s}$:

$$\tilde{h}_i - \tilde{h}_i^0 = \sum_j O_{ij} \frac{1}{\sqrt{\lambda^j}} s_j \tag{5.6}$$

It is easily seen that in the new variables the well becomes spherical: $\mathcal{G}(\mathbf{s}) \approx \mathcal{G}(\tilde{\mathbf{h}}0) + \frac{1}{2}\sum_i s_i^2$. The thermodynamic force $\partial \mathcal{G}/\partial s_i$ now reads

$$\frac{\partial \mathcal{G}}{\partial s_i} = \sum_j \frac{\partial \mathcal{G}}{\partial \tilde{h}_j} O_{ji} \frac{1}{\sqrt{\lambda^i}} \tag{5.7}$$

and can be easily calculated from expression (5.3).

The metadynamics equations in the new variables read

$$\mathbf{s}^{t+1} = \mathbf{s}^t + \delta s \frac{\boldsymbol{\phi}^t}{|\boldsymbol{\phi}^t|}$$

$$\mathcal{G}^t(\mathbf{s}) = \mathcal{G}(\mathbf{s}) + \sum_{t' < t} W e^{-\frac{|\mathbf{s}-\mathbf{s}^{t'}|^2}{2\delta s^2}} \tag{5.8}$$

where $\boldsymbol{\phi}^t = -\partial \mathcal{G}^t/\partial \mathbf{s}$. We follow here the prescription of Ref. [14] which relates the Gaussian height W to the Gaussian width δs, in the form $W \sim \delta s^2$. This prescription is similar to the one proposed in Ref. [11].

We now comment about the calculation of the Hessian matrix. In principle, it would be optimal to recalculate it and accordingly redefine the collective coordinates every time the system undergoes a transition to a new structure. However, the eigenvector corresponding to the largest eigenvalue is in most cases approximately parallel to the direction (111 000). Therefore a set of coordinates calculated for one structure might still be usable for simulation of a series of transitions since the separation between the direction of the volume gradient and the other degrees of freedom spanning the orthogonal subspace remains approximately preserved. When the bulk modulus of the system changes considerably which is often the case when the system changes coordination, the Hessian eigenvalues may also change substantially. In such a case it is preferred to recalculate the Hessian matrix and continue metadynamics with new collective coordinates suitable for the new initial structure.

5.4
Practical Aspects

Here we discuss some practical aspects related to the implementation and practical use of the technique. The advantage of the algorithm is that it can be very easily adapted for use with various MD codes. Since the exploration of the Gibbs free energy landscape is performed by means of a series of NVT simulations (an alternative continuous version of metadynamics is proposed in Ref. [21]), any MD code is suitable if it allows for simulation of a crystal in a general (triclinic) supercell and is able to calculate the stress tensor. The procedure can be performed by a simple code acting as a driver which writes the input files for the MD simulation, then calls the MD code executable via a system call and when the MD run is completed, extracts the average value of the stress tensor, reads the MD output files and writes new input files for the next run (Figure 5.1). The length of the MD run should be sufficient to allow both for a good equilibration of the system after change of the supercell induced by metadynamics and for a good averaging of the stress tensor. Even more important, however, is to allow sufficient time for the system to complete the transition, once this starts. Using too short simulation for one metastep might result in creation of defective or even disordered structures, since the structural transformation is disturbed by the rapidly changing supercell before it is complete.

In the case of classical potentials the above requirements do not pose a practical problem. For systems containing several hundreds of atoms an MD run of several thousand MD steps, representing several ps, takes a wallclock time of the order of 1 min. Since a typical metadynamics simulation takes of the order of hundred metasteps, it can be readily performed in few hours on a PC. The situation is different in the case of *ab initio* simulations. If the system is insulating and Γ point description is sufficient, for a typical system of about 100 atoms, running on a parallel computer with 16 cores, one ionic step of *ab initio* Born–Oppenheimer MD may take about 10 s. In such a case performing at each metastep an MD run of several thousand steps might be prohibitively expensive. It is possible to run metadynamics even performing at each metastep just few hundred MD steps and in such case one metastep might take just few hours. A typical simulation of hundred metasteps then could take about 10 days which is still feasible. In case of systems requiring the use of k-points the situation is clearly more complicated and here it is an advantage if the *ab initio* code is parallelized also over k-points and a sufficient number of computing cores are available.

Since during a structural transition the system might release a large amount of heat, it is convenient to use some kind of temperature control in order to prevent melting or creation of structural defects. To this end, simple velocity rescaling might be sufficient. If available in the MD code, more sophisticated thermostats such as Berendsen thermostat [22] or Nose-Hoover thermostat [23] can be used.

An important issue is the choice of the Gaussian parameters. Since at the beginning no information is available about the landscape of the Gibbs free energy of the system, such as the height of the barrier or the distance between the minima in the order parameter space, there is no obvious way to determine the appropriate values. Using the relation $W \sim \delta s^2$ mentioned in the previous section we need to specify only one parameter. A convenient way to find a starting value is based on the assumption that the barrier for the transition in the supercell is typically of the order of several 10 meV multiplied by the number of structural units N_u in

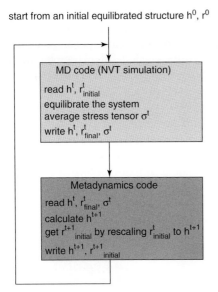

Figure 5.1 Flow chart of a metadynamics simulation showing the communication between the MD code and the metadynamics driver.

the system. From this assumption one can take the Gaussian height per structural unit, e.g., of the order of $\frac{W}{N_u} \sim 10$ meV. This represents a resolution in the free energy space corresponding to a fraction of the total barrier corresponding to the supercell. For illustration, in a system consisting of $N_u = 100$ structural units, the choice of a Gaussian width $\delta s = 40$ (kbar Å^3)$^{\frac{1}{2}}$ and corresponding Gaussian height $W = \delta s^2 = 1600$ kbar Å$^3 \approx 1$ eV corresponds to a resolution of $\frac{W}{N_u} \approx 10$ meV per structural unit and might be a good starting choice. Clearly this choice is highly dependent on the nature of the interactions in the system as well as on the simulation parameters, in particular pressure P. While at thermodynamic transition pressure the barrier might be large, upon increasing the pressure it becomes lower. The nature of the interactions determines the characteristic energy scale of structural transformations and clearly a crystal with strong covalent bonds and reconstructive transformations will have much higher barriers between the different phases than, e.g., a molecular crystal bound by van der Waals forces. If the Gaussian size is too large for a given system and conditions, metadynamics will not perform a proper filling of the initial free energy well but instead quickly escape from it in a random direction. Such situation is easily recognized by the fact that the first transition occurs within few metasteps. In such a case it is definitely necessary to reduce the Gaussian size in order to obtain a proper exploration of the landscape.

5.5
Examples of Applications

In this section we review several recent applications of the metadynamics technique to various kinds of crystals. The initial applications using a tight-binding scheme and classical force fields were reviewed in Refs. [15, 24]. More recently metadynamics studies using classical potentials were performed for MgSiO$_3$ enstatites in Refs. [25, 26]. These materials are important crustal and upper mantle minerals and the metadynamics studies succeeded in revealing microscopic details of complex transformation mechanisms of the ortho to high-pressure clinoenstatite transition [25] (Figure 5.2) and proto to high-pressure clinoenstatite transition [26]. Metadynamics with a classical force field has also been applied to the study of phase transitions in CdSe [27, 28]. In this system the forward transition from four-coordinated structures, wurtzite and zinc-blende, to six-coordinated rocksalt structure was studied. Besides that, the reverse transition induced by decompression of the rocksalt structure was studied and it was found that the final four-coordinated structure represents a mixture of layers with wurtzite and zinc blende stacking (Figure 5.3). This interesting outcome is caused by kinetic effects and is in agreement with a

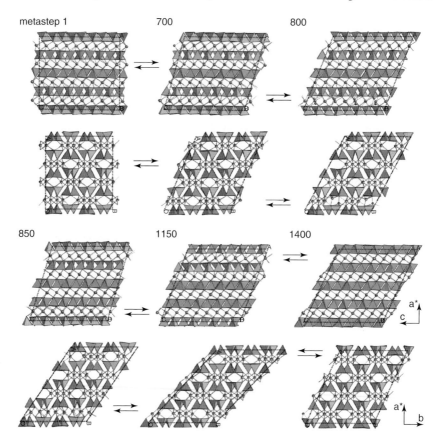

Figure 5.2 Selected relaxed structures from metadynamics simulation in projections along [010] (above) and [001] (below). The SiO$_4$ tetrahedra are represented by polyhedra and the Mg cations by spheres. a^* is the projection of a perpendicular to b and c. The position of the (100) slip planes and the respective slip directions are indicated by arrows. After Ref. [25].

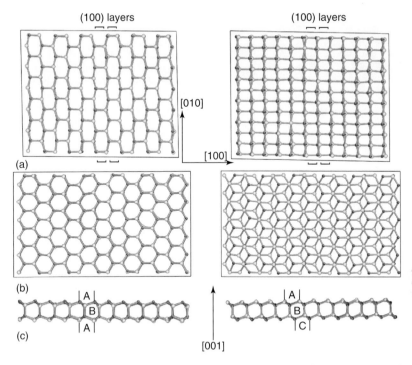

Figure 5.3 Left and right: snapshots of two different pairs of (001) layers in the 1728-atom CdSe sample during the transition from six- to four-coordination. (a) During metastep 353, alternating bonds have broken in adjacent (100) layers, such that all the four-membered rings have been converted into six-membered rings. In the pair of (001) layers shown on the left, the bonds broken in each (001) layer are the same, while in those shown on the right alternating bonds have been broken in each (001) layer. (b) During metastep 368, each (001) plane contains an array of hexagonal six-membered rings. In the two (001) planes on the left the six-membered rings are stacked directly above one-another ("eclipsed"), while in the planes shown on the right the stacking of the rings alternates ("staggered"). (c) The layers during metastep 368 shown along the (010) direction. A puckering of each of the (001) layers occurred. The eclipsed stacking of the six-membered rings combined with the puckering led to ABA WZ stacking, while the staggered stacking of six-membered rings gave the ABC ZB motif. After Ref. [27].

recent study [29] where both experiment and simulations using the transition path sampling [30] technique were performed.

In the following we focus on the applications of the technique where metadynamics was combined with density functional theory (DFT) calculations or neural-network representation of DFT calculations. The *ab initio* calculations in the case of $MgSiO_3$ and silica were performed with the Car–Parrinello method [31] using the CPMD code [32] and in the case of CO_2, C, and Ca with the projector augmented-wave (PAW) method [33] as implemented in the molecular-dynamics program VASP (Vienna ab-initio simulation program) [34, 35].

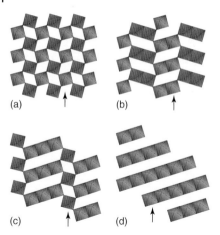

Figure 5.4 MgSiO$_3$ polytypes (after Ref. [18]): a – perovskite (space group *Pbnm*), d – postperovskite (*Cmcm*), b,c – newly found structures 2 × 2(*Pnma=Pbnm*) and 3 × 1(*P2$_1$/m*), respectively. Only silicate octahedra are shown; Mg atoms are omitted for clarity. In the postperovskite structure, the predicted plastic slip plane {110} is shown by an arrow. Arrows also show the likely slip planes in the other structures.

The first application of *ab initio* metadynamics was the study of the perovskite to postperovskite transition in MgSiO$_3$ [18], following the experimental discovery of the latter phase [36, 37]. This phase transition is of high importance for geophysics and seismology since it is related to the seismic anomaly in the D$''$ layer of the lowermost mantle. The perovskite structure was found to transform to the layered structure called postperovskite under conditions that correspond to those of the lowermost mantle (125–136 GPa, 2500–4000 K). The discovery of this transition, previously unknown, offered a possibility to explain anomalous properties of the D$''$ layer. The metadynamics simulations of this transition were performed using the original version of the algorithm [13] and revealed that the layered structure of postperovskite and the original perovskite structure are actually end members of an infinite series of polytypes and the transition mechanism involves plane sliding (Figure 5.4). Besides that it provided information about the dominant planes of plastic slip in this system, which is important for the texture and anisotropic properties of the D$''$ layer.

Metadynamics has also been applied to a number of polymorphs of silica. Silica is a material of high importance both for geophysics and for practical applications and it has a large number of polymorphs. Due to the fact that most phase transformations in this system are strongly reconstructive and require crossing of high barriers, their investigation poses a problem for both experiment and simulation. Kinetic effects play a strong role here, causing metastability of many phases as well as dependence of the structure on the history of the changes of pressure and temperature. We mention first the simulation of transformation of coesite which was performed using *ab initio* metadynamics [14, 38]. Experimentally coesite was found to amorphize upon compression at room temperature [39] and the same outcome was found in simulation [40]. *Ab initio* metadynamics was applied in order to see whether coesite under pressure could possibly transform also to another crystalline structure if amorphisation could be avoided. The simulation started from a 48-atom supercell of coesite and resulted in transformation to

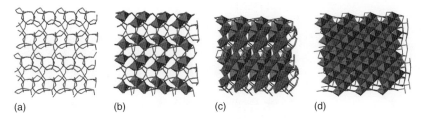

Figure 5.5 Structural evolution during the transition from coesite (a) to the α-PbO$_2$ phase (d). Intermediate states (b) and (c) show the initial growth and competition of chains of octahedra in different planes. After Ref. [14].

the metastable α-PbO$_2$ structure (Figure 5.5). This result represents a theoretical prediction since such a transition has not yet been observed experimentally.

Another silica structure that has been recently studied by *ab initio* metadynamics is α-cristobalite [41]. This material is another example of a system where strong kinetic effects cause the observation of an anomalous sequence of phases upon compression at room temperature. At pressures in the range 370–450 kbar, where at room temperature stishovite is the stable phase, a transition to poststishovite α-PbO$_2$ is instead observed [42, 43]. In Ref. [41] *ab initio* metadynamics was performed in order to elucidate the transformation mechanism leading to the metastable phase. Several simulations were performed and pathways leading to both α-PbO$_2$ structure and stishovite were found. In both cases the transformation mechanism involves several intermediate states. The enthalpy evolution in both cases is shown in Figure 5.6 and the structure evolution along the pathway leading to the α-PbO$_2$ structure is shown in Figure 5.7. In order to understand the preference for the α-PbO$_2$ structure observed in the experiment at room temperature the enthalpy profiles along each path were calculated between the initial structure and the first intermediate. It was found that the barrier leading to the α-PbO$_2$ structure is sligthly smaller which can explain the observed behavior.

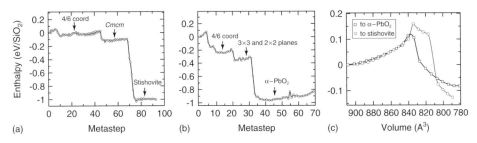

Figure 5.6 Evolution of the enthalpy during the transitions at 260 kbar and 600 K from cristobalite-XI to stishovite (a) and to α-PbO$_2$ (b). The activation barriers of the first step of the mechanisms (c) have been computed by optimizations of the atomic positions at the fixed h matrix. h values are determined by linear interpolation from cristobalite-XI to the first intermediates. After Ref. [41]. (Please find a color version of this figure on the color plates.)

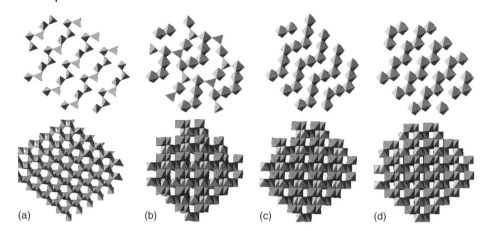

Figure 5.7 Section of a (11$\bar{2}$) plane (top row) and side view (bottom row) of the metastable structures encountered in the transition from cristobalite-XI (a) to α-PbO$_2$ (d). The transition occurs via the formation of a mixed tetrahedral and octahedral structure (b) and of a defective octahedral structure made of alternating 2 × 2 and 3 × 3 planes (c). After Ref. [41]. (Please find a color version of this figure on the color plates.)

Another important material that was simulated with metadynamics is silicon. Since in this case the high-pressure phases are metallic and their enthalpy differences are small, the proper DFT treatment would require a large number of k-points. This would make the simulation computationally highly demanding. For this reason in the study [44, 45] the recently developed neural network scheme [46] was applied which constructs an effective potential by means of a neural-network fitting scheme. The parameters of this scheme are fitted to the DFT calculated energies of a large number of solid, liquid, and amorphous structures. The resulting scheme combines the accuracy of DFT with the speed of classical force fields and therefore it was chosen for the metadynamics study of silicon. The study was started from the cubic diamond structure and succeeded in finding all stable structures of silicon up to the pressures of about 1 Mbar where silicon adopts close-packed structures. Upon increasing pressure silicon passes through the sequence of phases including cubic diamond, β-tin, *Imma*, simple hexagonal, *Cmca*, hcp, and fcc structures. This sequence was well reproduced by simulations which moreover revealed possible transformation mechanisms (Figures 5.8 and 5.9). In some of these mechanisms elements of nucleation can be seen as shown in Figure 5.8.

An interesting case of pressure-induced structural transformations is represented by pressure-induced polymerization where molecular crystal converts into one with an extended covalent network. An example of such a system is CO$_2$. Due to the chemical similarity between carbon and silicon one might expect that CO$_2$ could also exist in structures similar to those found in silica. Under normal conditions, however, CO$_2$ is gaseous. At low temperatures it becomes solid and creates a molecular crystal, the well-known dry ice (phase I). Upon increasing pressure other stable and metastable molecular phases are created (phases II and III). With

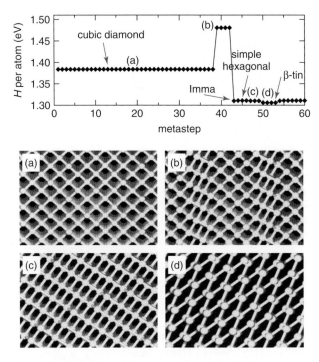

Figure 5.8 Enthalpies of the optimized crystal structures obtained in a metadynamics simulation at 300 K starting from the cubic diamond structure at an external pressure of 12 GPa. After Ref. [44].

further compression the intermolecular C–O distances continue to decrease and when these become comparable to the intramolecular ones, which occurs at the pressure of several 100 kbar, it becomes possible for each carbon atom to bound to four neighboring carbon atoms via bridging oxygens (similar to SiO_2). Such polymerization was studied experimentally in Refs. [47–51] (for review see Ref. [52]). Depending on the initial phase as well as pressure and temperature conditions various polymeric phases with crystalline (phases V [50] and VI [48]) and amorphous structure (Refs. [49, 51]) were found. It has been proposed that such phases could be superhard [50]. However, the experimental results are not conclusive and in some cases it was not possible to unambiguously resolve the X-ray diffraction pattern since the structures that were proposed as best fit are not mechanically stable at the DFT level of theory. In order to clarify the situation metadynamics was recently applied to study pressure-induced transitions from phases II and III [53]. Starting from phase II at 60 GPa and 600 K it was found that the system undergoes polymerization gradually, where in each step two adjacent molecular layers combine and create one layer of the polymeric structure (Figure 5.10). Depending on the degree of polymerization there is a variety of possible intermediate structures. A comparison of calculated X-ray diffraction pattern and Raman spectra to the experimental ones suggests that the phase VI found in the experiment is most likely a result

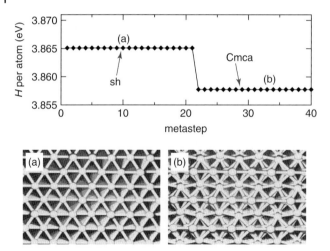

Figure 5.9 Enthalpies of the optimized crystal structures obtained in a metadynamics simulation at 800 K starting from the simple hexagonal (sh) structure at an external pressure of 45 GPa. After Ref. [44].

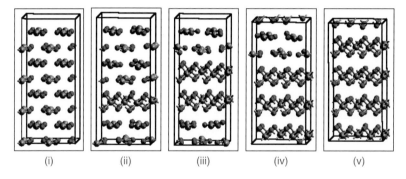

Figure 5.10 Structural evolution at steps 1, 4, 8, 86, and 89 in metadynamics simulations for a 32-molecule Phase II supercell of CO_2 at 60 GPa and 600 K. After Ref. [53]. (Please find a color version of this figure on the color plates.)

of incomplete transformation of phase II into a layered polymeric structure. The simulation starting from phase III was performed at 80 GPa and several temperatures between 0 and 700 K. It was found that the transformation is strongly temperature dependent and produces different outcomes above and below 300 K. Above 300 K an amorphous polymeric structure is produced which agrees with the experimental findings of Refs. [49, 51]. Below 300 K the resulting structure becomes more crystalline and at low temperatures (∼100 K) a perfect α-cristobalite-like structure (Figure 5.11) is produced. Since an experiment at high pressure and low temperature has not yet been performed, this result represents a prediction. We note here that by using the genetic algorithm USPEX [54] it was predicted that

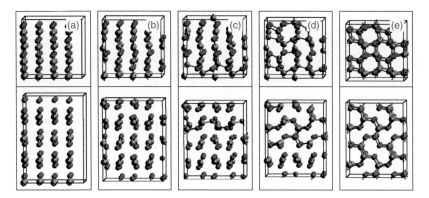

Figure 5.11 Illustration of the intermediate structures (metasteps 1, 14, 16, 17, 100) during the transformation of CO_2 from phase III (Cmca) to the α-cristobalite-like phase ($P4_12_12$) at 80 GPa and 300 K, where upper frames are top views and lower frames are side views. The density increases from 3.63 to 4.23 g/cm^3 from metastep 1 to metastep 100. After Ref. [53]. (Please find a color version of this figure on the color plates.)

the stable phase of CO_2 at pressures above 20 GPa is β-cristobalite [55] and at the simulation pressure of 80 GPa this phase has enthalpy lower by about 0.13 eV per CO_2 group with respect to α-cristobalite. The creation of the latter structure observed in the simulation therefore points to the importance of kinetic effects.

A particularly interesting problem is the study of postdiamond phases of carbon. This problem is of academic as well as of practical interest since diamond itself is used in high-pressure experiments to compress other materials and therefore its mechanical stability sets an upper limit to the pressure that can be achieved in the diamond-anvil cell experiments. Diamond is the hardest known mineral and is known to be stable in an extremely broad range of pressures – from about 5 GPa up to the largest pressures reached so far in the diamond anvil cell, which are of the order of 300 GPa. Very recently, due to the use of Ramp-wave compression technique it was possible to reach a peak pressure of 1400 GPa and measure the equation of state of diamond up to 800 GPa [56]. It was found that the diamond phase remains stable up to this pressure. Theoretically, the question of post diamond phases of carbon was dealt with since the advent of *ab initio* techniques. In Refs. [57, 58] it was shown that the diamond phase remains stable up to 1.1 TPa, where the more dense but still tetrahedrally coordinated BC8 phase becomes stable. At 2.9 TPa the BC8 phase transforms to the metallic simple cubic phase. The choice of the candidate phases was based on educated guess and only recently the BC8 phase was confirmed by genetic algorithms [4]. In dynamical simulations [59] using the PR technique, however, diamond was found to be stable at temperature of 1000 K up to about 3 TPa where it converted to a metastable metallic SC4 phase. In Ref. [60] metadynamics was applied to study transformation of carbon at extreme pressures. It was found that kinetic effects play a major role here and at 4000 K and 2 TPa cubic diamond transforms directly to the simple cubic carbon SC1 (*Pm-3 m*) (Figure 5.12), thus skipping the BC8 phase because of high

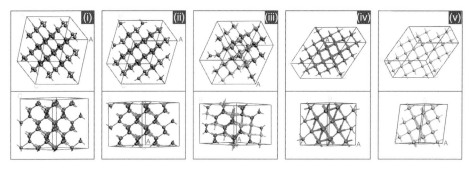

Figure 5.12 Metadynamics simulations of compressing cubic diamond at 2 TPa and 4000 K. Structural evolution of transformation from cubic diamond to SC1 (*Pm-3m*). After Ref. [60]. (Please find a color version of this figure on the color plates.)

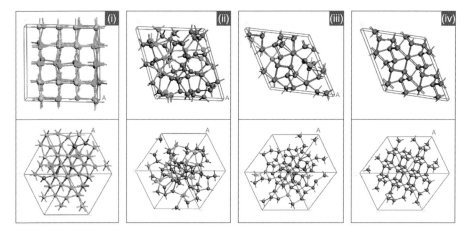

Figure 5.13 Structural evolution during metadynamics simulation yielding BC8 carbon (*Ia-3*) by decompressing SC1 (*Pm-3m*) carbon to 1 TPa at 5000 K. After Ref. [60]. (Please find a color version of this figure on the color plates.)

barrier. Upon decompression to 1 TPa at 5000 K the SC1 phase then transforms to the tetrahedral BC8 phase (Figure 5.13). The use of high temperatures appears to be crucial to obtain the stable phases since the compression of the cubic diamond phase at room temperature yielded the same metastable SC4 structure as found previously in Ref. [59] while decompression of the SC1 phase at 3000 K resulted in a new metastable tetrahedrally coordinated structure MP8. In this work the application of *ab initio* metadynamics not only succeeded in finding the previously proposed high-pressure BC8 and simple cubic phases of carbon directly from dynamical simulations but also allowed us to find kinetic pathways to these dense phases.

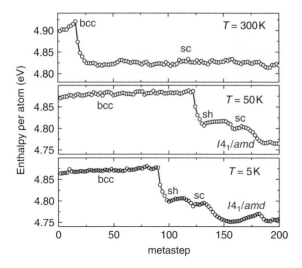

Figure 5.14 Evolution of the enthalpy starting with the initial bcc structure of Ca at 40 GPa at 300, 50, and 5 K. After Ref. [61].

To date the most recent application of *ab initio* metadynamics is the study of high-pressure phases of calcium [61]. This element has been extensively studied and among all elements has at high pressure the highest observed temperature of the superconducting transition. An unresolved issue is the discrepancy between the experimental observation of the simple cubic structure in the pressure range 32–119 GPa and room temperature [62] and DFT calculations which show that this structure is mechanically unstable [63, 64]. In Ref. [61] metadynamics as well as genetic algorithm [4, 54, 65] were applied to clarify this problem. Metadynamics runs at 40 GPa and low temperatures starting from the bcc structure (Figure 5.14) showed a sequence of transitions passing through intermediate simple hexagonal and simple cubic structures and finishing in a new $I4_1/amd$ structure which is actually isostructural with β-tin. The same structure was found with the genetic algorithm as the lowest enthalpy structure at 40 GPa and $T = 0$. These results suggest that the simple cubic structure at room temperature might be entropically stabilized and at low temperatures it is likely to transform to the $I4_1/amd$ phase. This also represents a new theoretical prediction to be verified by experiment.

5.6
Conclusions and Outlook

As demonstrated in a number of applications, the metadynamics-based technique for simulation of structural phase transitions in crystals effectively addresses severe timescale gap problems occurring in the variable cell constant-pressure MD simulations. The main achievements of the new method can be summarized as follows: (1) substantially improved predictive power – phases that would be skipped with previous techniques can be found now, (2) it takes into account entropic effects and can predict finite temperature phases stabilized by entropy, (3) it takes into account kinetic effects and thus can predict transitions into metastable phases, (4) the transitions can be simulated without excessive over-pressurization and therefore the method is able to provide a much more realistic microscopic transition mechanism, and (5) not only simple transitions "from A to B" but also complex transitions proceeding via several intermediate states can be studied. Due to these properties the approach represents a major step forward.

Clearly, there are some limitations of the method, both practical and conceptual. Concerning practical difficulties, while *ab initio* metadynamics on parallel computers is nowadays well feasible, the problem may appear in the case of systems requiring large number of *k*-points where the simulation could become computationally very expensive.

The conceptual limitations are essentially threefold. First, although the supercell order parameter works well in a number of crystalline systems, it cannot be expected to be universal order parameter. In particular, it might not work in the case of organic crystals consisting of molecules with internal degrees of freedom (e.g., torsions). In such a case the metadynamics scheme has to be adapted accordingly in order to include the degrees of freedom driving the transition. Second, even though in some cases elements of nucleation are observed in the transition process, the use of a relatively small supercell with periodic boundary conditions together with the collective order parameter still to some extent favors a collective mechanism of the transition. Proper simulation of nucleation in structural phase transitions is an open problem and progress in this direction has been achieved with the application of the transition path sampling technique [30], as shown, e.g., in Refs. [29, 66].

Third, simulations of structural transitions in real systems have to take into account the role of defects, particularly the extended ones, such as dislocations. These are likely to act as nucleation centers and their proper inclusion would require simulation of substantially larger systems. At present this might be possible only with classical potentials and therefore besides the timescale problem one also encounters a length-scale problem here. We conclude by saying that fully realistic simulations of structural transitions in real crystalline systems still remain a challenge.

Acknowledgments

The author is deeply grateful to all collaborators who participated in this work over several years, namely (in alphabetical order): C. Bealing, J. Behler, M. Bernasconi, C. Ceriani, D. Donadio, S. Jahn, D.D. Klug, A. Laio, M.S. Lee, C. Molteni, J. Montoya, A. R. Oganov, M. Parrinello, P. Raiteri, S. Scandolo, J. Sun, E. Tosatti, Y. Yao, and F. Zipoli. It was a pleasure to work with them.

The author has been supported by the Slovak Research and Development Agency under Contracts No. APVV-0442-07 and No. VVCE-0058-07 and by the Vega Project No. 1/0096/08.

References

1. Schön, J.C. and Jansen, M., (1996) First step towards planning of syntheses in solid-state chemistry: Determination of promising structure candidates by global optimization. *Angew. Chem. Int. Ed.*, **35**, 1287–1304.
2. Pickard, C.J. and Needs, R.J., (2006) High-pressure phases of silane. *Phys. Rev. Lett.*, **97**, 045504.
3. Bush, T.S., Catlow, C.R.A., and Battle, P.D., (1995) Evolutionary programming techniques for predicting inorganic crystal structures. *J. Mater. Chem.*, **5**, 1269–1272.
4. Oganov, A.R. and Glass, C.W., (2006) Crystal structure prediction using evolutionary algorithms: principles and applications. *J. Chem. Phys.*, **124**, 244704.
5. Woodley, S.M., Battle, P.D., Gale, J.D., and Catlow, C.R.A., (1999) The prediction of inorganic crystal structures using a genetic algorithm and energy minimization. *Phys. Chem. Chem. Phys.*, **1**, 2535–2542.
6. Parrinello, M. and Rahman, A., (1980) Crystal structure and pair potentials: A molecular dynamics study. *Phys. Rev. Lett.*, **45**, 1196–1199.
7. Parrinello, M. and Rahman, A., (1981) Polymorphic transitions in single crystals: A new molecular dynamics method. *J. Appl. Phys.*, **52**, 7182–7190.
8. Focher, P., Chiarotti, G.L., Bernasconi, M., Tosatti, E., and Parrinello, M., (1994) Structural phase transformations via first-principles simulation. *Europhys. Lett.*, **26**, 345–351.
9. Laio, A. and Parrinello, M., (2002) Escaping free-energy minima. *Proc. Natl. Acad. Sci. USA*, **99**, 12562–12566.
10. Huber, T., Torda, A.E., and van Gunsteren, W.F., (1994) Local elevation: A method for improving the searching properties of molecular dynamics simulation. *J. Comput.-Aided Mol. Des.*, **8**, 695–708.
11. Grubmüller, H., (1995) Predicting slow structural transitions in macromolecular systems: Conformational flooding. *Phys. Rev. E*, **52**(3), 2893–2906.
12. Voter, A.F., (1997) Hyperdynamics: Accelerated molecular dynamics of infrequent events. *Phys. Rev. Lett.*, **78**, 3908–3911.
13. Martoňák, R., Laio, A., and Parrinello, M., (2003) Predicting crystal structures: The Parrinello–Rahman method revisited. *Phys. Rev. Lett.*, **90**, 075503.
14. Martoňák, R., Donadio, D., Oganov, A.R., and Parrinello, M., (2006) Crystal structure transformations in SiO_2 from classical and ab initio metadynamics. *Nature Mater.*, **5**, 623–626.
15. Martoňák, R., Laio, A., Bernasconi, M., Ceriani, C., Raiteri, P., Zipoli, F., and Parrinello, M., (2005) Simulation of structural phase transitions by metadynamics. *Zeitschrift für Kristallographie*, **220**, 489–498.
16. Ceriani, C., Laio, A., Fois, E., Gamba, A., Martoňák, R., and Parrinello, M., (2004) Molecular dynamics simulation of reconstructive phase transitions on anhydrous Li-ABW zeolite. *Phys. Rev. B*, **70**, 113403.
17. Ishikawa, T., Nagara, H., Kusakabe, K., and Suzuki, N., (2006) Determining the structure of phosphorus in phase IV. *Phys. Rev. Lett.*, **96**(9), 095502.
18. Oganov, A.R., Martoňák, R., Laio, A., Raiteri, P., and Parrinello, M., (2005) Anisotropy of earth's D'' layer and stacking faults in the $MgSiO_3$ post-perovskite phase. *Nature*, **438**, 1142–1144.
19. Quigley, D. and Probert, M.I.J., (2005) Phase behavior of a three-dimensional core-softened model system. *Phys. Rev. E (Statistical, Nonlinear, and Soft Matter Phys.)*, **71**(6), 065701.
20. Raiteri, P., Martoňák, R., and Parrinello, M., (2005) Exploring polymorphism: the case of benzene. *Angew. Chem. Int. Ed.*, **44**, 3769–3773.

21. Pagliai, M., Iannuzzi, M., Cardini, G., Parrinello, M., and Schettino, V., (2006) Lithium hydroxide phase transition under high pressure: An *ab initio* molecular dynamics study. *Chemphyschem*, **7**, 141–147.
22. Berendsen, H.J.C., Postma, J.P.M., van Gunsteren, W.F., DiNola, A., and Haak, J.R., (1984) Molecular dynamics with coupling to an external bath. *J. Chem. Phys.*, **81**, 3684.
23. Nosé, S., (1984) A molecular dynamics method for simulations in the canonical ensemble. *Mol. Phys.*, **52**, 255–268.
24. Martoňák, R., Oganov, A.R., and Glass, C.W., (2007) Crystal structure prediction and simulations of structural transformations: metadynamics and evolutionary algorithms. *Phase Transit.*, **80**, 277–298.
25. Jahn, S. and Martoňák, R., (2008) Plastic deformation of orthoenstatite and the ortho to high-pressure clinoenstatite transition: A metadynamics simulation study. *Phys. Chem. Minerals*, **35**, 17–23.
26. Jahn, S. and Martoňák, R., (2009) Phase behavior of protoenstatite at high pressure studied by atomistic simulations. *Am. Mineral.*, **94**, 950–956.
27. Bealing, C., Martoňák, R., and Molteni, C., (2009) Pressure-induced structural phase transitions in CdSe: A metadynamics study. *J. Chem. Phys.*, **130**, 124712.
28. Bealing, C., Martoňák, R., and Molteni, C., (2010) The wurtzite to rock salt transition in CdSe: A comparison between molecular dynamics and metadynamics simulations. *Solid State Sci.*, **12**, 157–162.
29. Leoni, S., Ramlau, R., Meier, K., Schmidt, M., and Schwarz, U., (2008) Nanodomain fragmentation and local rearrangements in CdSe under pressure. *PNAS*, **105**, 19612–19616.
30. Dellago, C., Bolhuis, P.G., Csajka, F.S., and Chandler, D., (1998) Transition path sampling and the calculation of rate constants. *J. Chem. Phys.*, **108**, 1964–1977.
31. Car, R. and Parrinello, M., (1985) Unified approach for molecular dynamics and density-functional theory. *Phys. Rev. Lett.*, **55**, 2471–2474.
32. CPMD V3.9 Copyright IBM Corp 1990–2004, Copyright MPI fuer Festkoerperforschung Stuttgart 1997–2001.
33. Blöchl, P.E., (1994) Projector augmented-wave method. *Phys. Rev. B*, **50**, 17953–17979.
34. Kresse, G. and Furthmüller, J., (1996) Efficient iterative schemes for ab initio total-energy calculations using a plane-wave basis set. *Phys. Rev. B*, **54**, 11169–11186.
35. Kresse, G. and Joubert, D., (1999) From ultrasoft pseudopotentials to the projector augmented-wave method. *Phys. Rev. B*, **59**, 1758–1775.
36. Murakami, M., Hirose, K., Kawamura, K., Sata, N., and Ohishi, Y., (2004) Post-perovskite phase transition in $MgSiO_3$. *Science*, **304**, 855–858.
37. Oganov, A.R. and Ono, S., (2004) Theoretical and experimental evidence for a post-perovskite phase of $MgSiO_3$ in earth's D'' layer. *Nature*, **430**, 445–448.
38. Martoňák, R., Donadio, D., Oganov, A.R., and Parrinello, M., (2007) From four- to six-coordinated silica: transformation pathways from metadynamics. *Phys. Rev. B*, **76**, 014120.
39. Hemley, R.J., Jephcoat, A.P., Mao, H.K., Ming, L.C., and Manghnani, M.H., (1988) Pressure-induced amorphization of crystalline silica. *Nature*, **334**, 52–54.
40. Dean, D.W., Wentzcovitch, R.M., Keskar, N., Chelikowsky, J.R., and Binggeli, N., (2000) Pressure-induced amorphization in crystalline silica: Soft phonon modes and shear instabilities in coesite. *Phys. Rev. B*, **61**, 3303–3309.
41. Donadio, D., Martoňák, R., Raiteri, P., and Parrinello, M., (2008) The influence of temperature and anisotropic pressure on the phase transitions of α-cristobalite. *Phys. Rev. Lett.*, **100**, 165502.
42. Dubrovinsky, L.S., Dubrovinskaia, N.A., Saxena, S.K., Tutti, F., Rekhi, S., Le Bihan, T., Shen, G., and Hu, J., (2001) Pressure-induced transformations of cristobalite. *Chem. Phys. Lett.*, **333**, 264–270.
43. Shieh, S.R., Duffy, T.S., and Shen, G., (2005) X-ray diffraction study of phase

stability in SiO$_2$ at deep mantle conditions. *Earth Planet. Sci. Lett.*, **235**, 273–282.
44. Behler, J., Martoňák, R., Donadio, D., and Parrinello, M., (2008) Metadynamics simulations of the high-pressure phases of silicon employing a high-dimensional neural network potential. *Phys. Rev. Lett.*, **100**, 185501.
45. Behler, J., Martoňák, R., Donadio, D., and Parrinello, M., (2008) Pressure-induced phase transitions in silicon studied by neural network-based metadynamics simulations. *physica status solidi (b)*, **245**, 2618–2629.
46. Behler, J. and Parrinello, M., (2007) Generalized neural-network representation of high-dimensional potential-energy surfaces. *Phys. Rev. Lett.*, **98**, 146401.
47. Iota, V., Yoo, C.S., and Cynn, H., (1999) Quartzlike carbon dioxide: An optically nonlinear extended solid at high pressures and temperatures. *Science*, **283**, 1510–1513.
48. Iota, V., Yoo, C., Klepeis, J., Jenei, Z., Evans, W., and Cynn, H., (2007) Six-fold coordinated carbon dioxide VI. *Nature Materials*, **6**, 34–38.
49. Kume, T., Ohya, Y., Nagata, M., Sasaki, S., and Shimizu, H., (2007) A transformation of carbon dioxide to nonmolecular solid at room temperature and high pressure. *J. Appl. Phys.*, **102**, 53501–53505.
50. Yoo, C.S., Cynn, H., Gygi, F., Galli, G., Iota, V., Nicol, M., Carlson, S., Häusermann, D., and Mailhiot, C., (1999) Crystal structure of carbon dioxide at high pressure: "superhard" polymeric carbon dioxide. *Phys. Rev. Lett.*, **83**, 5527.
51. Santoro, M., Gorelli, F.A., Bini, R., Ruocco, G., Scandolo, S., and Crichton, W.A., (2006) Amorphous silica-like carbon dioxide. *Nature*, **441**, 857.
52. Santoro, M. and Gorelli, F.A., (2006) High pressure solid state chemistry of carbon dioxide. *Chemical Society Reviews*, **35**, 918–931.
53. Sun, J., Klug, D.D., Martoňák, R., Montoya, J.A., Lee, M.-S., Scandolo, S., and Tosatti, E., (2009) High-pressure polymeric phases of carbon dioxide. *PNAS*, **106**, 6077–6081.
54. Glass, C.W., Oganov, A.R., and Hansen, N., (2006) Uspex – evolutionary crystal structure prediction. *Comp. Phys. Comm.*, **175**, 713–720.
55. Oganov, A.R., Ono, S., Ma, Y.M., Glass, C.W., and Garcia, A., (2008) Novel high-pressure structures of MgCO$_3$, CaCO$_3$, and CO$_2$ and their role in Earth's lower mantle. *Earth and Planet. Sci. Lett.*, **273**, 38–47.
56. Bradley, D.K., Eggert, J.H., Smith, R.F., Prisbrey, S.T., Hicks, D.G., Braun, D.G., Biener, J., Hamza, A.V., Rudd, R.E., and Collins, G.W., (2009) Diamond at 800 GPa. *Phys. Rev. Lett.*, **102**, 075503.
57. Yin, M.T., (1984) Si-III (bc-8) crystal phase of Si and C: Structural properties, phase stabilities, and phase transitions. *Phys. Rev. B*, **30**, 1773–1776.
58. Yin, M.T. and Cohen, M.L., (1983) Will diamond transform under megabar pressures? *Phys. Rev. Lett.*, **50**, 2006–2009.
59. Scandolo, S., Chiarotti, G.L., and Tosatti, E., (1996) Sc4: A metallic phase of carbon at terapascal pressures. *Phys. Rev. B*, **53**, 5051–5054.
60. Sun, J., Klug, D.D., and Martoňák, R., (2009) Structural transformations in carbon under extreme pressure: Beyond diamond. *J. Chem. Phys.*, **130**, 194512.
61. Yao, Y., Klug, D.D., Sun, J., and Martoňák, R., (2009) Structural prediction and phase transformation mechanisms in calcium at high pressure. *Phys. Rev. Lett.*, **103**, 055503.
62. Gu, Q.F., Krauss, G., Grin, Yu., and Steurer, W., (2009) Experimental confirmation of the stability and chemical bonding analysis of the high-pressure phases Ca-I, II, and III at pressures up to 52 GPa. *Phys. Rev. B*, **79**, 134121.
63. Gao, G., Xie, Y., Cui, T., Ma, Y., Zhang, L., and Zou, G., (2008) Electronic structures, lattice dynamics, and electron–phonon coupling of simple cubic Ca under pressure. *Solid State Commun.*, **146**, 181–185.
64. Teweldeberhan, A.M. and Bonev, S.A., (2008) High-pressure phases of calcium and their finite-temperature phase boundaries. *Phys. Rev. B*, **78**, 140101(R).

65. Yao, Y., Tse, J.S., and Tanaka, K., (2008) Metastable high-pressure single-bonded phases of nitrogen predicted via genetic algorithm. *Phys. Rev. B*, **77**, 052103.
66. Zahn, D. and Leoni, S., (2004) Nucleation and growth in pressure-induced phase transitions from molecular dynamics simulations: Mechanism of the reconstructive transformation of NaCl to the CsCl-type structure. *Phys. Rev. Lett.*, **92**, 250201.

6
Global Optimization with the Minima Hopping Method
Stefan Goedecker

6.1
Posing the Problem

Configurations of molecules and other condensed matter systems are stable at zero temperature if the forces acting on the nuclei vanish and if small displacements away from the configuration are giving rise to a force that is pushing the system back to the original configuration. These requirements are fulfilled for all the local minima of the potential energy surface which is typically denoted as the Born–Oppenheimer surface if it is calculated by solving the electronic Schrödinger equation. The potential energy surface gives the energy $E(\mathbf{R}_1, \mathbf{R}_2, \ldots, \mathbf{R}_{Nat})$ of a system as a function of all the atomic coordinates $\mathbf{R}_i$ of a condensed matter system being composed of Nat atoms. At finite temperature the atoms oscillate in the catchment basin [1] of a local minima and the system can jump into the catchment basin of another possibly lower local minimum if the barrier separating the two catchment basins is not too high. Local minima thus become metastable configurations whose lifetime depends on the barrier heights of the surrounding saddle points and on the temperature. Being able to find the global minimum and other local minima is a very important task since it allows one to predict the structure of matter. The structure in turn determines all other properties of the system.

Finding a local minimum is in principle an easy task. One can just move downhill in the direction opposite to the gradient and if one proceeds by sufficiently small steps, one is guaranteed to end up in a local minimum. This procedure corresponds to the steepest descent minimization method which is rather slow in practice but conceptually very simple. Finding the global minimum is a much more difficult task. There is no simple prescription which is guaranteed to find the global minimum. Even worse, there is not even a criterion that would allow one to determine whether one has found the global minimum or just a low-energy local minimum.

Perhaps the simplest idea to find the global minimum would be to perform local geometry optimizations starting from many different random starting structures. This approach is doomed to fail except for very small systems since the number of

Modern Methods of Crystal Structure Prediction. Edited by Artem R. Oganov
Copyright © 2011 WILEY-VCH Verlag GmbH & Co. KGaA, Weinheim
ISBN: 978-3-527-40939-6

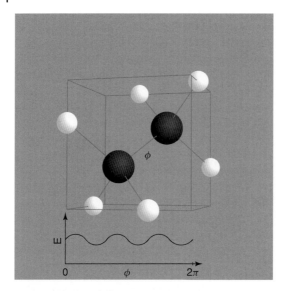

Figure 6.1 The two monomer alkane C_2H_6 which has three local minima. The carbon atoms are represented by black spheres and the hydrogen atoms by white spheres.

local minima increases exponentially with system size. For large systems, one can therefore never locate the majority of the local minima to determine subsequently which one is the lowest in energy. The exponential growth of the number of local minima can be easily seen for two representative systems. As a representative for large molecule systems and in particular for large biomolecules let us consider the alkane family C_nH_{2n+2}. Figure 6.1 shows the shortest polymer C_2H_6. If one performs a rotation by 120° of the upper CH_3 unit around the axis connecting the two C atoms one obviously ends up with an equivalent configuration. The energy of the C_2H_6 as a function of the rotation angle ϕ therefore has a periodicity of three as shown in the inset and as a consequence three minima. Hence the C_2H_6 has at least three local minima. As more C atoms are added to build up longer alkane chains, one can do rotations around all the bonds connecting neighboring carbon atoms and one thus expects of the order of 3^{n-1} local minima.

This exponential increase in the number of local minima is not a special property of chain molecules. For silicon one arrives at the same estimate. Figure 6.2 shows the Wooten Wear process [2] that allows us to create amorphous silicon structures from the perfect crystal. Of the order of 3^{Nat} amorphous structures can be created in this way for a cell containing Nat silicon atoms.

Another simple approach for finding the global minimum is the thermodynamic approach, which alleviates the problem of the enormous number of local minima. Since one is, in general, only interested in the energetically lowest ones, one could try to generate a Boltzmann distribution by the Monte Carlo method. In this Boltzmann distribution the lowest energy structures have the highest weight. Hence they are the most likely to be found. The lower the temperature the higher

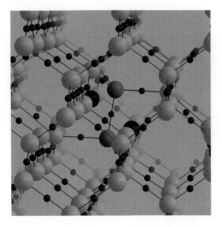

Figure 6.2 The green spheres show the silicon atoms in a perfect silicon crystal. The little blue dots represent the electrons which form chemical bonds. Each silicon atom has four bonds with its four nearest neighbors. Two silicon atoms were moved from the perfect crystal positions, pictured by the black spheres, to the new positions indicated by the two red spheres. The new system has again fourfold coordination and is a local minimum of the potential energy surface [3]. By repeated moves of this type, one can obtain crystalline structures with more and more point defects which finally become amorphous structures. (Please find a color version of this figure on the color plates.)

is the weight of the global minimum with respect to the weights of the other structures. This approach has two drawbacks. For small temperatures the simulation gets trapped in most cases in some region of the configuration space which does not contain the global minimum. This happens because all trajectories that lead into the global minimum have to cross high-energy regions. Configurations in these high-energy regions are rejected in the Metropolis step of the Monte Carlo method, thus preventing the crossing. Barriers between the catchment basins of neighboring local minima can be eliminated if the true potential energy surface is replaced by a transformed potential energy surface. The energy of this transformed surface is constant in each catchment basin and is given by the value of the local minimum in the catchment basin. This transformed surface is implicitly the surface in all methods using local geometry optimizations such as the Monte Carlo plus local optimization method [4]. Unfortunately this approach does not eliminate barriers between different funnels as shown in Figure 6.3. As a consequence the temperature has to be sufficiently large to allow crossings of high-energy regions between low-energy funnels. This implies that the Boltzmann weight for the higher energy configurations becomes larger and that one has to visit again a rather large number of minima before the global minimum is included with high probability in the sample. The optimal temperature which is the best compromise between the conflicting requirements of a large weight for low-energy configurations and the absence of trapping can in general not be predicted.

There is a second problem with the thermodynamic approach. By construction it gives a distribution. This means that one will find the global minimum again

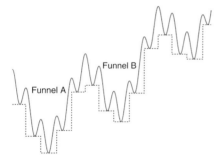

Figure 6.3 The pictures show a potential energy surface (full line) together with the transformed surface (dotted line), which is constant within each catchment basin and equal to its value at the minimum. Even though the transformed surface has not any more barriers between neighboring catchment basins, it has still barriers between funnels.

and again if one does a very long run which is not trapped in a wrong funnel. However, in the context of global optimization the form of the distribution is not of interest and it would be enough to find each local minimum just once and then either stop or explore other minima.

6.2
The Minima Hopping Algorithm

The minima hopping algorithm [5] is based on two basic principles. By a built-in feedback mechanism trapping is excluded. It recognizes when regions that were already visited previously during a run are revisited. In this case it makes more violent moves which will force it to explore different regions of the configurational space. Secondly, it exploits the Bell–Evans–Polanyi principle [6] for the moves from one catchment basin into another one. Exploiting this principle gives a higher probability that the catchment basin into which one hops belongs to a low-energy local minimum.

The Bell–Evans–Polanyi principle is a well-known empirical observation that strongly exothermal chemical reactions have typically low activation energies. As shown in Figure 6.4 it can be derived within a very simple model where we have two local minima along a reaction coordinate. The whole potential energy surface along the reaction coordinate is given by the pieces of two parabolas centered in the two local minima. The intersection of the two parabolas where the potential energy surface switches from one parabola to the other is then the barrier of this chemical reaction where the system starts from the educt represented by the left local minimum and then goes over the barrier into the product which is represented by the right local minimum. The reaction is more exothermic if the product minimum is low in energy. If the product minimum is lowered, the right-hand parabola is shifted down and the intersection between the two parabolas is also lowered. Hence the barrier height for the chemical reaction as well as the activation energy decreases.

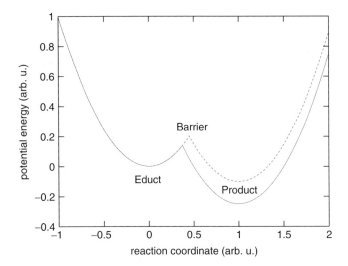

Figure 6.4 Illustration of the Bell–Evans–Polanyi principle as described in the text. The dashed curve gives one potential energy surface and the full curve another one where the parabola of the product has been shifted down.

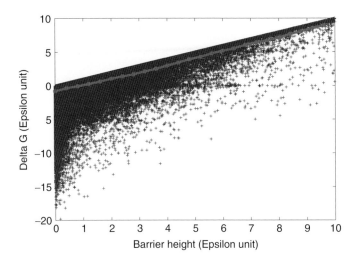

Figure 6.5 The relation between the barrier height and the energy difference ΔG between the two local minima for 130 000 saddle points in a 55 atom Lennard–Jones cluster. The red curve is the average over all the data. This curve shows a linear relation between the barrier height and the energy difference between the minima and thus validates the Bell–Evans–Polanyi principle on average even though some data point fall far away from this curve.

This model is of course a crude simplification of a true potential energy surface. It neglects, for instance, the fact that the curvature of the parabola might change when it is shifted up or down. Hence it is not surprising that the Bell–Evans–Polanyi principle is not strictly followed by each pair of minima connected by a saddle point if one examines a large number of such pairs as shown in Figure 6.5. The same figure shows however also that it is surprisingly well followed on average. Such a validity on average is all that is needed in the context of a minima hopping global geometry optimizations since one is doing a large number of hops from one minimum to another.

A flowchart of the minima hopping algorithm is given below. As the name indicates, the algorithm jumps from one local minima to the next. It consists of an outer part where newly found minima are either accepted or rejected and an inner part where escape trials from the current local minimum are done. Let us first discuss this inner part.

```
            initialize a current minimum 'Mcurrent'

MDstart
        ESCAPE   TRIAL PART
            start a MD trajectory with kinetic energy Ekin from
            the current minimum 'Mcurrent' in a soft direction.
            Once potential energy reaches the mdmin-th minimum
            along the trajectory stop  MD and optimize geometry
            to find the closest local minimum 'M'

            if ('M' equals 'Mcurrent') then
                Ekin = Ekin*beta_same   (beta_same > 1)
                goto MDstart
            else if ('M' equals a minimum visited previously) then
                Ekin = Ekin*beta_old   (beta_old > 1)
            else if ('M' equals  a new minimum ) then
                Ekin = Ekin*beta_new   (beta_new < 1)
            endif

        DOWNWARD PREFERENCE PART
            if ( energy('M') - energy('Mcurrent') < Ediff ) then
                accept new minimum: 'Mcurrent' = 'M'
                add 'Mcurrent' to history list
                Ediff = Ediff*alpha_acc (alpha_acc < 1)
            else if rejected
                Ediff = Ediff*alpha_rej (alpha_rej > 1)
            endif

        goto MDstart
```

From the current local minimum M_{cur} we start a short molecular dynamics (MD) trajectory with initial velocities which are scaled such that we obtain a certain kinetic energy E_{kin}. During the MD part we monitor the potential energy as a function of time and the MD trajectory is stopped at the md_{min}th minimum of the potential energy. At such a minimum of the potential energy along the trajectory the system is of course not exactly in a minimum of the potential energy surface but it is in general close to such a minimum. The stopping condition is typically encountered after less than 100 MD steps. The final MD configuration is then used as a starting point for a standard local geometry optimization, which then brings the system into the closest local minimum M. Three scenarios are now possible. The MD trajectory has just performed some oscillations within the catchment basin of M_{cur} and hence M equals M_{cur}. In this case the value of E_{kin} is increased by the factor β_{same} and another escape trial is performed. This increase prevents that the system can get trapped in the catchment basin of M_{cur}. If the MD trajectory has crossed a barrier the minimum M will be different from M_{cur}. If it is a minimum that was visited before the kinetic energy is increased by β_{old}, otherwise it is decreased by β_{new}. The motivation for increasing it is that we do not want to revisit known regions of the configuration space. By using higher kinetic energies, we can overcome higher barriers into unexplored regions. The motivation for decreasing E_{kin} is the Bell–Evans–Polanyi principle. We want to cross the lowest possible barriers because it is more likely to find a low-energy local minimum behind a low-energy barrier than behind a high barrier. The examination of the Bell–Evans–Polanyi principle above was for the case where one moves along the reaction coordinate, i.e., where one crosses from one catchment basin into another one exactly at the saddle point. The MD trajectories used in the minima hopping method do not cross exactly at the saddle point. Nevertheless the Bell–Evans–Polanyi principle remain also valid in the case where MD trajectories are used to cross from one catchment basin into another one. This was examined in a detailed way for Lennard–Jones and silicon clusters [7] and is also in agreement with our experience for all other systems we have examined.

If one would use really small kinetic energies for an ordinary MD trajectory within the minima hopping algorithm, one would try to observe a rare event, namely the crossing of a barrier. This would be inefficient since one would have to follow the MD trajectory over many oscillations within the current catchment basin before it crosses into another catchment basin. This problem can be avoided if one does not use random directions for the velocity vector at the start of MD trajectory as one would do in a standard MD simulations but if one uses a velocity vector which points in soft directions, i.e., in a direction in which the curvature of the potential energy is small. Along such a direction the probability is high that one will find a low-energy barrier after a small number of MD steps. Starting in such a soft direction means that high frequency oscillations are virtually absent in the MD trajectory which then allows us to use a larger time step. Finding a soft direction can be done at the price of a few force evaluations with the dimer method [8]. The effect of an MD trajectory along a soft direction is illustrated in Figure 6.6. One sees that within a short MD trajectory, global movements are occurring whereas in a standard MD simulation, one can observe only local oscillations of the atoms without any significant global movement [9]. So a movement over small barriers does not cause small modifications in the structure but on the contrary rather large modifications.

The outer part of the minima hopping algorithm is a standard acceptance/rejection step which can either be based on thresholding or on a Metropolis step. In the case of thresholding a new minimum is accepted if its energy E is not higher than the energy of the current minimum E_{cur} by E_{diff}. In the case of

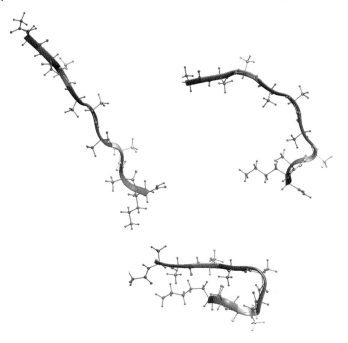

Figure 6.6 Three snapshots of an MD trajectory of polyalanin along a soft direction. (Please find a color version of this figure on the color plates.)

a Metropolis step the new minimum is always accepted if its energy is lower ($E < E_{cur}$) and with probability $\exp(-(E - E_{cur})/E_{diff})$ if its energy is higher. The parameter E_{diff} is adjusted continuously during the simulations. It is increased by a factor $\alpha_{rej} > 1$ if the new minimum is rejected and decreased by a factor $\alpha_{acc} < 1$ if it is accepted. By choosing $\alpha_{rej} = 1/\alpha_{acc}$ half of all the minima are accepted on average during a minima hopping run. The thresholding version is more efficient for structure seekers where it is rather easy to fall down to the global minimum funnel, whereas in multifunnel systems the Metropolis version can be somewhat more efficient because it allows to jump earlier out of a wrong funnel which does not contain the global minimum. Even if the Metropolis version is used minima hopping does not create a thermodynamic distribution, because E_{diff} is permanently modified to adapt it to the part of configuration space that is currently being searched over.

The performance of minima hopping is rather insensitive to the exact values of α_{acc}, α_{rej}, β_{same}, β_{old} and β_{new} as shown in Table 6.1. Setting $\alpha_{acc} = 1/\alpha_{rej}$ and $\beta_{same} = \beta_{old} = 1/\beta_{new}$ ensures that the number of local geometry optimizations is about four times the number of accepted local minima. In this case the number of local geometry optimizations needed is four times the number of accepted local minima. What is very important is that the parameters satisfy the explosion condition. This condition will finally lead to an explosion or fragmentation of a

Table 6.1 The influence of the choice of the available free parameters on the performance of minima hopping[a].

β_{same}	β_{old}	β_{new}	α_{acc}	α_{rej}	n_{soft}	md_{min}	n_{min}
1.05	1.12	1/1.12	1/1.12	1.12	20	2	237
1.05	1.12	1/1.12	1/1.12	1.12	25	2	251
1.05	1.12	1/1.12	1/1.12	1.12	25	3	290
1.05	1.12	1/1.12	1/1.12	1.12	25	1	279
1.05	1.15	1/1.15	1/1.10	1.10	20	2	244
1.05	1.15	1/1.15	1/1.10	1.10	20	3	259
1.05	1.15	1/1.15	1/1.10	1.10	15	2	250
1.05	1.15	1/1.15	1/1.10	1.10	10	2	275
1.05	1.20	1/1.20	1/1.10	1.10	15	2	280
1.05	1.05	1/1.05	1/1.05	1.05	20	4	321
1.01	1.15	1/1.15	1/1.10	1.10	20	2	275
1.03	1.15	1/1.15	1/1.10	1.10	20	2	243
1.05	1.15	1/1.15	1/1.15	1.15	20	2	293
1.05	1.10	1/1.10	1/1.10	1.10	20	2	267
1.10	1.10	1/1.10	1/1.10	1.10	20	2	272
1.05	1.10	1/1.10	1/1.10	1.10	20	2	277
1.05	1.05	1/1.05	1/1.05	1.05	20	2	240

[a] The results were obtained for the Lennard–Jones 38 cluster by averaging over 100 runs. The initial positions were random.

cluster being simulated in a very long minima hopping run. This is due to the fact that once the majority of the minima have been visited, the kinetic energy is increased all the time. Due to the Bell–Evans–Polanyi principle the MD trajectories end up in higher energy local minima and therefore E_{diff} has to go up as well. Once the global minimum is found, the minima hopping algorithm thus explores higher and higher energy regions until finally the system explodes. This explosion is not an important aspect of the method in practice. However, what is important in practice is that the explosion condition also allows the algorithm to rapidly leave funnels and thus leads us to high efficiency. The explosion condition is derived by excluding the possibility that all the high-energy local minima are rejected while a certain subset of low-energy minima is accepted all the time.

Let us denote by N_{same}, N_{old}, and N_{new} the number of minima that are classified as "same," "old," and "new," respectively, during a minima hopping run and by N_{acc}, and N_{rej}, respectively, the number of accepted and rejected minima. If E_{kin} and E_{diff} do not tend to infinity or to zero the conditions

$$\beta_{same}^{N_{same}} \beta_{same}^{N_{old}} \beta_{new}^{N_{new}} \approx 1 \tag{6.1}$$

and

$$\alpha_{acc}^{N_{acc}} \alpha_{rej}^{N_{rej}} \approx 1 \tag{6.2}$$

have to be satisfied. We can now classify the events in a more detailed way by distinguishing both between acceptance/rejectance and same/old/new. We introduce the variables N_{same}, $N_{\text{old,acc}}$, $N_{\text{new,acc}}$, $N_{\text{old,rej}}$ and $N_{\text{new,rej}}$, which count the number of times these events happen. The algorithm gets trapped if no new minima are accepted, i.e., if $N_{\text{new,acc}} = 0$ and no old minima are rejected $N_{\text{old,rej}} = 0$. To prevent this, we request that E_{kin} explodes, i.e., increases exponentially in this case and Eq. (6.1) becomes

$$\beta_{\text{same}}^{N_{\text{same}}} \beta_{\text{same}}^{N_{\text{old,acc}}} \beta_{\text{new,rej}}^{N_{\text{new}}} > 1 \tag{6.3}$$

Since $\beta_{\text{same}}^{N_{\text{same}}}$ is greater than 1, the above condition is fulfilled if

$$\beta_{\text{same}}^{N_{\text{old,acc}}} \beta_{\text{new,rej}}^{N_{\text{new}}} \geq 1 \tag{6.4}$$

The ratio of $N_{\text{old,acc}}$ to $N_{\text{new,rej}}$ can be obtained under these circumstances from Eq. (6.2) which becomes (replacing $\approx$ by $=$).

$$\alpha_{\text{acc}}^{N_{\text{old,acc}}} \alpha_{\text{rej}}^{N_{\text{new,rej}}} = 1 \tag{6.5}$$

Inserting this ratio into Eq. (6.4) gives the final explosion condition [10]

$$\frac{\log(\alpha_{\text{rej}})}{\log(1/\alpha_{\text{acc}})} \geq \frac{\log(1/\beta_{\text{new}})}{\log(\beta_{\text{old}})} \tag{6.6}$$

Because we want to have small kinetic energies due to the Bell–Evans–Polanyi principle, we generally choose the smallest value of β_{old} which satisfies the above condition and leads to an equality in the above equation. If the explosion condition is violated the system could get trapped or, what occurs more frequently in practice, the system spends just a long time in a certain region of the configurational space and the search is very inefficient.

The effect of the feedback is illustrated in Figure 6.7, which shows the history of a minima hopping run to find the global minimum of a 512-atom NaCl cluster [11, 12]. Ionic clusters are systems which find their global ground state rather easily. They try to adopt cubic structures [13, 14] and the 512-atom cluster has therefore a magic size. In addition funnels can rather easily be visualized. The global minimum funnel is around the $8 \times 8 \times 8$ cubic structure. Other funnels correspond to flawed orthorhombic structures of different side length such as a $7 \times 8 \times 9$ structure. From Figure 6.7, one sees that the global minimum is found after having searched over a little bit more than 2000 configurations. During this first part the kinetic energy of the MD trajectory E_{kin} oscillates around the value 0.8 mHa. Once the global minimum has been found more and more configurations in the global minimum funnel are revisited and the kinetic energy increases rapidly to a peak value of 2.5 mHa which allows the system then to escape from the global minimum funnel. After the escape the system cools down rapidly, i.e., E_{kin} decreases and the system falls into the $7 \times 8 \times 9$ funnel. Then the temperature rises again slightly to allow the system to escape from the $7 \times 8 \times 9$ funnel and to jump a second time into the global minimum funnel. In the last part of the simulation three more such cycles are shown where the temperature rises to allow escapes from funnels and the global minimum is in total found five times.

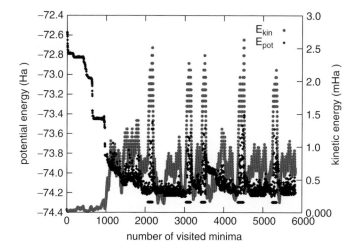

Figure 6.7 The history of a minima hopping global optimization of a 512-atom NaCl cluster starting from random initial positions. The black dots show the energies of the various local minima which are visited during the simulation. The grey dots show the evolution of the quantity E_{kin}. The behavior of E_{diff} is not shown since it tracks the behavior of E_{kin}. The similar behavior of both quantities is due to the Bell–Evans–Polanyi principle.

In the above description of this process we have associated E_{kin} with a temperature since the temperature of an MD trajectory is proportional to its kinetic energy. If E_{kin} is converted to a temperature it turns out that one obtis values which are of the order of the melting temperature. If the temperature rises due to the feedback mechanism, one can therefore say that the system undergoes some kind of melting which allows it to escape from a funnel.

The influence of the performance on all the parameters of the minima hopping algorithm described so far are summarized in Table 6.1. The influence of the parameter n_{soft} is the most important. n_{soft} gives the number of force evaluations used in the dimer method to find a soft initial direction for the MD trajectory. Larger values of n_{soft} give softer directions which allow in turn to find with a short MD trajectory low-energy barriers. Too much softening has however to be avoided because in this case all the moves from a given local minimum would be in the same direction of the softest mode and the ergodicity would be lost. Another important parameter is md_{min}. md_{min} is a measure of the length of the MD trajectory. It counts over how many potential energy minima along the trajectory one continues the MD part before stopping it. 50 force evaluations were needed on average for each MD escape trial. The influence of the parameters α_{acc}, α_{rej}, β_{same}, β_{old} and β_{new} is rather weak as long as they are close to one and as long as the explosion condition is fulfilled. Hence we use in many applications simply 1.05 for all parameters which are larger than one and 1/1.05 for all parameters smaller than one. The numbers in Table 6.1 are considerably better than the numbers given in the original paper [5], where softening in the MD part was not yet used. The softening allows to reduce the average E_{kin} by about a factor of 2. The numbers are also better than the numbers in Ref. [15] where no effort was made to tune the parameters and at least 1190 local geometry optimization were needed to find the global minimum. In the best runs of this table less than 1000 geometry optimizations were needed.

The minima hopping method has been tested for the Lennard–Jones clusters with up to 150 atoms which are a standard benchmark system for global optimization algorithms. It was able to find all the known global minima including difficult cases such as LJ_{38} and LJ_{75}. This is in contrast to standard genetic algorithms which have problems finding nonicosahedral ground states [15, 16]. As a test, we also applied minima hopping to the LJ_{1000} cluster for which the putative global minimum was found by a lattice based method that uses some structural knowledge [17]. Starting from random positions minima hopping was also able in this case to find the global minimum after having searched over a few million configurations.

6.3
Applications of the Minima Hopping Method

One appealing feature of the minima hopping method is that it can be applied without major modifications to any system. This is due to the fact that molecular dynamics and local geometry optimizations can be done for virtually any system. One of the first applications of the minima hopping method was the structure determination of silicon clusters [18, 19]. This was one of the first applications where a systematic global search algorithm was used within density functional theory. This work showed that there is a large number of different structures that are all very close in energy. In many cases the energetic separation between the ground state and the first local minimum was just a few $k_B T$ at room temperature. So the Boltzmann distribution does not give a unique ground state at room temperature. In addition, it is most difficult to give a correct energetic ordering for the various structures if their energy differences are so small. This work also showed that it is very dangerous to use less accurate descriptions for such clusters. With a good force field scheme one has for instance typically errors in the total energy of a few electron volts. Within this energy interval there are thousands of force field configurations and it would be very inefficient to consider them all as candidates for the global minimum in subsequent density functional geometry optimization. More recently we have also shown that the potential energy surface for silicon systems given by density functional theory is much smoother than the potential energy surface given by more approximate schemes [20]. Hence a much smaller number of local minima has to be visited in a density functional based minima hopping simulations for finding the global minimum compared to the case where more approximate description schemes for silicon are used. So even though an energy and force evaluation is much more costly in a density functional calculation than in a calculation using a force field, the overall computing time for reliably finding the global minimum is typically less if the global geometry optimization is done within density functional theory from the beginning.

We have also systematically studied gold clusters with up to 300 atoms [21]. This size is considerably larger than the sizes of gold clusters studied previously. As in the case of the silicon clusters it was found that there exist many different structures which are very close in energy to the ground state. In the case of the gold clusters, the energetically similar structures had frequently different structural motifs such as fcc-like structures and structures with a fivefold symmetry. Even though the total energy is very similar, the different contributions to the total energy are quite different. In the case of fcc-like structures the atoms in the core of the cluster are low in energy, but there is a large number of energetically unfavorable surface atoms. In the case of fivefold structures the surface is energetically more favorable but the atoms in the core are stressed and are therefore higher in energy [22]. Because these structures are so close in energy their energetic ordering can be changed by the addition of a single atom. This means that there are no structural windows within which one has a certain structural motif. These gold clusters are thus a nice illustration of the slogan of nanosciences that every single atom counts and that a single atom can change the structure and properties completely.

Minima hopping has also been used to create tip structures for silicon tips used in atomic force microscopy (AFM) [23]. It was found that realistic tip structures are rather floppy and jump in a dissipative way from one minimum into another one due to their interaction with the surface. In this way the experimentally observed dissipation could be explained for the first time.

Minima hopping has also been applied to the problem of protein folding [24]. In some cases the correct native state can be obtained whereas in other cases the structures obtained from the simulation are lower in energy than the experimental structures. This suggests that standard atomic force fields are not accurate enough to predict the small energy differences between different conformations of proteins.

Minima hopping can also be used to find low-energy defects. This is due to the fact that it explores higher and higher energy structures once the global minimum has been found. We used this feature to study defect structures in the B_{80} fullerene and found a rich variety of different defect structures [25].

6.4
Conclusions

Even though global optimization is a very important topic, not only in structure prediction but also in many other fields, mathematics does not give us efficient and generally applicable algorithms to solve it. The majority of the work of practitioners has a dominating empirical component.

The genetic algorithms with their huge number of empirically inspired flavors are a prominent example in this context. The minima hopping has also empirical components, but it introduces nevertheless two basic principles. The Bell–Evans–Polanyi principle that predicts which moves are good on average in a global optimization and the explosion condition which rigorously excludes trapping of the algorithm. In connection with a stable and accurate density functional method [26] minima hopping can be used to find global minima structures within the highly accurate density functional scheme. Minima hopping has up to now mainly been applied to the structure determination of nonperiodic systems, but it should be equally well applicable to periodic systems.

References

1. Wales, D. (2003) *Energy Landscapes* Cambridge University Press, Cambridge.
2. Wooten, F., Winer, K., and Weaire, D. (1985) Computer generation of structural models of amorphous Si and Ge. *Phys. Rev. Lett*, **54**, 1392.
3. Goedecker, S., Deutsch, T., and Billard, L. (2002) A fourfold coordinated point defect in silicon, *Phys. Rev. Lett*, **88**, 235501.
4. Li, Z. and Sheraga, H.A. (1987) Monte Carlo minimization approach to the multiple minima problem in protein folding. *Proc. Natl. Acad. Sci. USA*, **84**, 6611.
5. Goedecker, S. (2004) *J. Chem. Phys*, **120**, 9911; Goedecker, S., Hellmann, W., and Lenosky, T. (2005) *Phys. Rev. Lett*, **95**, 055501.
6. Jensen, F. (1999) *Computational Chemistry*, Wiley, New York.
7. Roy, S. (2008) Goedecker, S., and Hellmann, V. A Bell–Evans–Polanyi principle for molecular dynamics trajectories and its implications for global optimization, *Phys. Rev. E*, **77**, 056707.
8. Henkelman, G., and Jonsson, H. (1999) A dimer method for finding saddle points on high dimensional potential surfaces using only first derivatives. *J. Chem. Phys*, **111**, 7010.
9. Roy, S. (2009) PhD thesis, University of Basel.
10. The condition given in the original publication [5] is only valid for values of $\beta_2, \beta_3, \alpha_1, \alpha_2$ which are close to one in which case the equations agree to first order.
11. Amsler, M. (2009) Master thesis, University of Basel.
12. Amsler, M., Ghasemi, S. A., Goedecker, S., Neelov, A., and Genovese, L. (2009) Adsorption of small NaCl clusters on surfaces of silicon nanostructures. *Nanotechnology*, **20**, 445301.
13. Rose, J. P., and Berry, R. S. (1993) $(KCl)_{32}$ and the possibilities for glassy clusters, *J. Chem. Phys*, **98**, 3262.
14. Martin, T. P. (1980) The structure of ionic clusters: Thermodynamic functions, energy surfaces, and SIMS. *J. Chem. Phys*, **72**, 3506.
15. Schoenborn, S., Goedecker, S., Roy, S., and Oganov, A. R. (2009) Evolutionary algorithms and minima hopping for cluster structure prediction. *J. Chem. Phys*, **130**, 144108.
16. Hartke, B. J. (1999) *J. Comput. Chem*, **20**, 1752.
17. Yang, X., Cai, W., and Shao, X. (2007) *J. Comp. Chem*, **28**, 1427.
18. Goedecker, S., Hellmann, W., and Lenosky, T. (2005) Global minimum determination of the Born–Oppenheimer surface within density functional theory. *Phys. Rev. Lett*, **95**, 055501.
19. Hellmann, W., Hennig, R. G., Goedecker, S., Umrigar, C. J., Delley, B., and Lenosky, T. (2007) Questioning the existence of a well defined ground state for silicon clusters, *Phys. Rev. B*, **75**, 085411.
20. Ghasemi, S. A., Amsler, M., Hennig, R. G., Roy, S., Goedecker, S., Umrigar, C. J., Genovese, L., Lenosky, T. J., Morishita, T., and Nishio, K. The energy landscape of silicon systems and its description by force fields, tight binding schemes, density functional methods and quantum Monte Carlo methods. arXiv:0910.4050.
21. Bao, K., Goedecker, S., Koga, K., Lancon, F., and Neelov, A. (2009) Structure of large gold clusters obtained by global optimization using the minima hopping method. *Phys. Rev. B*, **79**, 041405.
22. (2005) Structural properties of nanoclusters: Energetic, thermodynamic, and kinetic effects. Francesca Baletto and

Riccardo Ferrando, *Rev. Mod. Phys*, **77**, 371.

23. Ghasemi, A., Goedecker, S., Lenosky, T., Hug, H., Meier, E., and Baratoff, A. (2008) Ubiquitous mechanisms of energy dissipation in noncontact atomic force microscopy. *Phys. Rev. Lett*, **100**, 236106.

24. Roy, S., Goedecker, S., Field, M. J., and Penev, E. (2009) A minima-hopping study of all-atom protein folding and structure prediction, *J. Phys. Chem. B*, **113** (2), pp. 7315–7321.

25. Bao, K., Goedecker, S., Genovese, L., Neelov, A., Ghasemi, S. A., and Deutsch, T. Structural stability of the B_{80} fullerene against defect formation. arXiv:0902.1599.

26. Genovese, L., Neelov, A., Goedecker, S., Deutsch, T., Ghasemi, A., Zilberberg, O., Bergman, A., Rayson, M., and Schneider, R. (2008) Daubechies wavelets as a basis set for density functional pseudopotential calculations. *J. Chem. Phys*, **129**, 014109.

7
Crystal Structure Prediction Using Evolutionary Approach
Andriy O. Lyakhov, Artem R. Oganov, and Mario Valle

If a mathematician had to formulate a problem that we face in crystal structure prediction, it would probably sound like this: "Find the global minimum on a very noisy landscape in a multidimensional space." The search space is so huge that one could not do an exhaustive search in any reasonable time even using all the computing power available for the mankind in a near future. This obviously nontrivial problem is further complicated by the fact that we do not know the exact form of the landscape. All we can do is to calculate the "height" (free energy) given the coordinates (structure parameters, such as atom positions) and relax the system to the nearest local minimum. This problem is in fact so hard that 15 years ago the answer on the question "Are the structures predictable?" was a clear ≪No≫: just by writing this concise statement, in what would be the first one-word paper in the chemical literature, one could safely summarize the present state of affairs" [1]; see also [2]. However mathematicians have been trying to solve similar problems for over half a century – since first computers were created. And actually they found a plethora of methods that are able to provide us with reasonably good solutions. Some of these methods are problem specific and their success is based on problem constrains and symmetries. But some approaches are very general and thus could be applied to crystal structure prediction (some of them are described in other chapters of this book). One of the most general and powerful techniques that at the same time is adaptable for any specific problem is the evolutionary algorithm approach.

Before going deeper into the details, we would like to mention that no method is currently available which is able to solve a *general* global optimization problem or even prove that the solution found by any means is in general case a global optimum. All algorithms are essentially heuristic – they are able to found a reasonable solution in a reasonable time without a guarantee that this solution is the best one or even close to it. We can only hope that it is. However, the situation is not as grim as one may think. Years of practice show us that usually solutions are quite good and quite often you can indeed find the best solution or a really close one. Evolutionary algorithms are not an exception. They were used in many different fields and produced great results where other methods failed. For example, an evolutionary approach could give a reasonably good solution for the

Modern Methods of Crystal Structure Prediction. Edited by Artem R. Oganov
Copyright © 2011 WILEY-VCH Verlag GmbH & Co. KGaA, Weinheim
ISBN: 978-3-527-40939-6

traveling salesman problem with 10 000 cities [3], where the number of possible solution candidates is far beyond astronomical ($\sim 10^{35\,000}$).

7.1
Theory

The name of the approach – evolutionary – indicates that it uses mechanisms inspired by biological evolution: reproduction, mutation, recombination, and selection. The candidate solutions to the optimization problem are individuals in a population that evolves under the repeated application of the above mechanisms, and a fitness function determines how much any of them survives. In more detail, the basic steps of the evolutionary technique (see Figure 7.1) are as follows:

1) The evolutionary approach starts by choosing the adequate representation for the problem: a one-to-one correspondence between the point in the search space and a set of numbers. This is a very important yet neglected step and the quality of the representation has direct impact on the effectiveness of the algorithm.

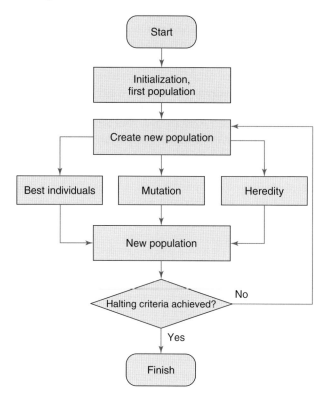

Figure 7.1 Block scheme of the typical evolutionary algorithm.

2) Initialization of the first generation, that is, a set of points in the search space that satisfy the problem constrains.
3) Determination of the quality for each member of the population using the so-called fitness function.
4) Selection of the "best" members from the current generation as parents from which the algorithm creates new points (offspring) in the search space by applying specially designed variation operators to them.
5) Evaluation of the quality for each new member of the population.
6) Selection of the "best" offspring to build the new generation of the population.
7) Repeat steps 4–6 till some halting criteria is achieved.

One can see that some biological terms are used to describe this approach:

- *Population.* A set of points in the search space (which we will further call solutions) that are analyzed as possible candidates for the optimal solution.
- *Parents.* A set of solutions that are used to create new candidates for the optimum solution.
- *Offspring.* A set of solutions created from parents using variation operators.
- *Selection.* A process that divides solutions into ones that have to "die" and ones that have to "survive" to build the next generation.

Two main classes of variation operators also have biological names:

- Heredity operators use a few parent solutions to build one offspring solution.
- Mutation operators use a single-parent solution to produce a single child.

Why are evolutionary algorithms so effective? No one actually knows, and as we wrote earlier no one could prove that the best solution found by the algorithm is in general close to the optimum one. However there are two things that inspired people to develop this class of algorithms and made us believe that this approach is indeed a good one. First of all in a correctly designed evolutionary algorithm the quality of the best solution(s) in each new generation (quantified by fitness function) is at least not worse than in the previous one. We can simply keep the best solution or any given number of best solutions, if none of the offspring solutions are better than them. And, secondly, you could see the success of the algorithm executed by Mother Nature if you look into the mirror or hug your pet.

There is one thing to keep in mind during the design and execution of every evolutionary algorithm. You have to deal with the trade-off: diversity of the population versus convergence to the optimum solution. Higher diversity of the population means that you can explore your search space better. However it slows down the process of finding the minimum even if you have a few structures in its basin of attraction. On the other hand, using elitist approaches and reducing the diversity helps your population collapse into a local minimum faster while the risk of omitting the global optimum is higher. The desired balance between diversity and convergence can usually be achieved only if you can tune it "on the fly" when you reveal more information about the system and your search space.

7.1.1
Search Space, Population, and Fitness Function

Crystal structure prediction requires us to find the global minimum on the free-energy landscape (that we will call the search space) for system of a given stoichiometry. Each point on this landscape (solution) represents a crystal structure with certain atomic positions and lattice vectors. The set of locally optimized solutions we will call a population.

One of the features of evolutionary algorithms, which is very helpful for the crystal structure prediction problem, is their ability to find metastable states – good local minima on the energy landscape that are clearly separated from the global minimum.

Fitness function describes the quality of each solution and allows us to compare them. Naturally, the free energy would be the relevant fitness function for a crystal structure prediction algorithm. Lower free energy will correspond to better solution and the most stable structure under given conditions will have the lowest fitness function value.

7.1.2
Representation

As we have already mentioned, the choice of the right representation is crucial for the effectiveness of the algorithm. One of the reasons why first attempts to design an evolutionary algorithm for crystal structure prediction ([4, 5]; see also [6]) had small success was counterproductive representation choice. In these approaches, a discrete grid of atom positions was used. The structural variables of the crystal were represented by a binary string, and standard evolution operators for binary strings were applied. These operators were not physically meaningful and, therefore, the algorithm was basically performing a random walk in the search space. Obviously it could reliably find the global minimum only in the simplest systems, and even in this case the number of optimizations it has to do is comparable with the random search methods. Later algorithms use the real-number representation for atom positions and lattice parameters [7, 8]. This representation requires more sophisticated variation operators that are better "optimized" for their task and allow researchers to build powerful methods for crystal structure prediction.

In the rest of this chapter, we will describe our evolutionary algorithm Universal Structure Predictor: Evolutionary Xtallography (USPEX) and results that were achieved with it [8, 9]. USPEX represents the coordinates of atoms in the unit cell and lattice vectors by real numbers. Therefore, the search space is continuous and not discrete like in algorithms with binary string representation. The difficulty of the problem is increased, but this is more than compensated by the possibility to construct physically intuitive and powerful variation operators. Using real number representation, we also have less risk to omit some shallow local minima.

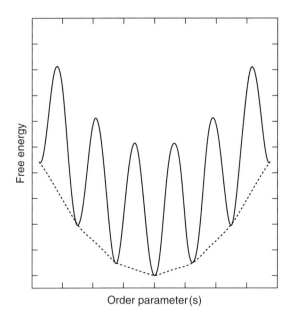

Figure 7.2 Reducing the noise in the complex search space after local optimization.

7.1.3
Local Optimization and Constrains

Each structure produced by USPEX and similar modern algorithms is locally optimized. Local optimization is a crucial part of the technique because otherwise it would be impossible to find the lowest free-energy structure for any reasonable complex system. It also makes the search space much more "smooth" (see Figure 7.2) and reduces the intrinsic dimensionality of the energy landscape,[1] see Table 7.1.

Since complexity of the problem has exponential dependence on the dimensionality, local optimization greatly simplifies the task of finding the global minimum on the energy landscape. USPEX uses external tools for local optimization. So far it can use VASP [12] and SIESTA [13] for first-principles optimization and GULP [14] for semiclassical simulations. It is also relatively easy to add support of other energy-computing engines, if needed.

We would like to mention that not every structure on the energy landscape is a chemically and physically feasible solution. If we try, for example, to optimize the structure where two atoms are located in the same spot then the calculation may

---

1) Intrinsic dimensionality is the minimum number of variables that are needed to represent the search space. Local optimization, by creating correlations between atomic positions (such as favoring the formation of some bonds and disfavoring others), decreases the intrinsic dimensionality and dramatically simplifies the crystal structure prediction problem.

Table 7.1 Intrinsic vs. extrinsic dimensionality[a].

System	Extrinsic dimensionality	Intrinsic dimensionality
Au_8Pd_4	39	10.9
$Mg_{16}O_{16}$	99	11.6
$Mg_4N_4H_4$	39	32.5

[a]The numbers were produced using the molecular visualization toolkit STM4 [10] using the Grassberger–Procaccia algorithm with a suitably adapted Camastra sampling correction [11].

"explode" the unit cell and we would not get any meaningful results. Therefore, we have to apply constrains to discard unfeasible solutions. The structure is considered unfeasible if:

1) The distance between any two atoms is smaller than threshold determined by user (e.g., there are no known bonds shorter than 0.5 Å). One can set different thresholds for different pairs of atom types separately; for example, the sum of correspondent atom radii.
2) One of the lattice vectors is too small. User can determine the threshold value; for example, it can be set to the diameter of the largest atom present in the system.
3) The angle between two lattice vectors is too small or the angle between the lattice vector and the diagonal of the parallelogram formed by other two lattice vectors is too small, see Figure 7.3. One can always choose the lattice vectors in such a way that the angle between any of them is in the $(60°, 120°)$ range. To do this, one can replace the longer vector (let's say **a**) in a pair violating the constrain by $\mathbf{a}' = \mathbf{a} - \operatorname{ceil}\left(|\mathbf{a} \cdot \mathbf{b}|/\|\mathbf{b}\|^2\right) \cdot \operatorname{sign}(\mathbf{a} \cdot \mathbf{b}) \cdot \mathbf{b}$ see [20] for more details. Our experience shows that this procedure speeds up local optimization and improves the efficiency of the algorithm as a whole[2].

7.1.4
Initialization of the First Generation

Common way to create the first generation in evolutionary approach is uniform random sampling, aimed at achieving high diversity of the population. In fact, if you do not have any *a priory* information about how the possible optimal solution could look like, random sampling is the most reasonable way to do the unbiased

---

2) The 60–120 degree conditions are a simplified version of the cell constraints – more complete conditions are that a unit cell vector (and also cell diagonals) should not have projections onto other cell vectors that are greater by absolute value than half of the latter vector [20].

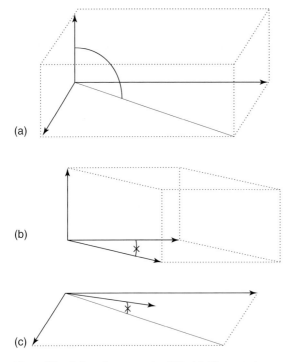

Figure 7.3 Cell angles constrains [20]. (a) All constrains satisfied, (b) angle between lattice vectors is too small, and (c) angle between lattice vector and the diagonal of the parallelogram formed by other two lattice vectors is too small.

initialization. You just pick up random points in the search space and check them for feasibility. Feasible solutions are added to the first generation. This process is repeated until the desired number of trial solutions is reached. In our case it means that we randomly choose the lattice vectors and then randomly drop the atoms into the unit cell. If some information about the optimum solution is known, for example, unit cell volume or lattice parameters, this can be used as constraints for optimization. In this case one would, for example, fix the lattice parameters and vary only atomic positions within the unit cell. User can also "seed" the first generation with the structures that seem reasonable (e.g., those known for similar compounds, or for the same compound at different conditions, or coming from previous structure prediction runs) and fill the rest of it with random ones.

High diversity of the first generation is the key to the success of the algorithm. If we do not have a structure in the basin of attraction (so-called funnel, see Figure 7.4) of the global minimum, then the probability to find that optimal solution can be low.

Random initialization, however, poses a problem relevant for large systems. When the number of atoms in the unit cell rises, randomly generated structures become more and more similar [15] from chemical point of view to each other

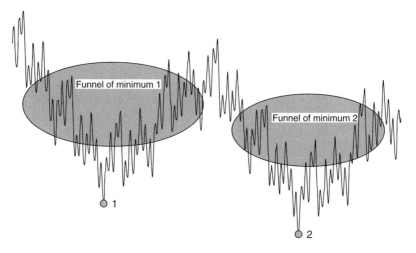

Figure 7.4 Basin of attraction (funnel) of a good local minimum is a set of surrounding local minima. Different funnels typically contain very distinct types of structures and are separated by high-energy barriers.

and to a disordered system. This can be visualized as trying to build a crystal by replicating small volumes of liquid. Thus purely random sampling cannot be used for effective crystal structure prediction in this case. However, there is a good method that combines relatively high diversity of the first population with a high degree of randomness. We call it unit cell splitting: large cell is split into smaller subcells that are filled with atoms randomly and then replicated to fill the full cell, see Figure 7.5a. Such structures have a higher translational symmetry and are more ordered, usually leading to a more diverse population. To help to break this

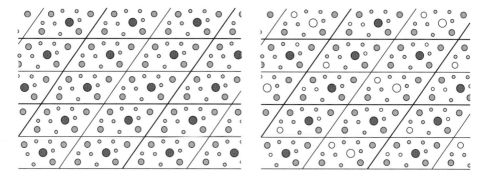

Figure 7.5 Schematic illustration of the (a) subcell and (b) pseudo-subcell splitting for compositions (a) $A_4B_{16}C_{20}$ and (b) $A_3B_{14}C_{16}$ (atoms A – large black circles, B – medium dark-gray circles, C – small gray circles, empty circles – vacancies). Thick lines show the true unit cell, split into four (pseudo-)subcells.

undesired "additional" symmetry by variation operators (described below) as well as further increase the diversity of the first population, different structures have to be split into different number of subcells.

For cells where the number of atoms (e.g., a prime number) is not good for splitting into a small number of identical subcells, the algorithm creates random vacancies to keep the correct number of atoms in the whole unit cell, see Figure 7.5b. In this case, no additional symmetry is induced and nontrivial solutions can be found. We also believe that this pseudo-subcell method is able to improve conventional random sampling methods [16, 17] when dealing with large systems.

7.1.5
Variation Operators

In general, the choice of variation operators follows naturally from the representation. Mutation operators usually randomly distort the numbers from the set that represents the solution, while heredity operators combine different parts of these sets from different parent solutions into one child solution. For real-number crystal structure representation, we use two different types of mutation operators – lattice mutation and atom permutation.

Lattice mutation applies strain matrix with zero-mean Gaussian random strains to the lattice vectors:

$$L'_i = (I + \varepsilon)L_i = \begin{pmatrix} 1+\varepsilon_1 & \varepsilon_6/2 & \varepsilon_5/2 \\ \varepsilon_6/2 & 1+\varepsilon_2 & \varepsilon_4/2 \\ \varepsilon_5/2 & \varepsilon_4/2 & 1+\varepsilon_3 \end{pmatrix} L_i$$

Lattice mutation is shown in Figure 7.6. The position of atoms (their fractional coordinates within the lattice) remains unchanged. This operator allows the algorithm

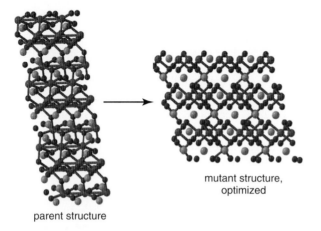

mutant structure, optimized

parent structure

Figure 7.6 Lattice mutation applies strain to the lattice vectors. From Oganov et al. [9].

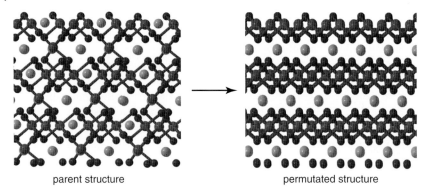

Figure 7.7 Permutation operator swaps identities of a few atom pairs. From Oganov et al. [9].

to investigate the neighborhood of good individuals. Also sometimes structures of similar quality differ essentially in lattice and that is why the premature convergence of the lattice to some "optimal" value reduces the effectiveness of the algorithm. Lattice mutation operator increases the diversity of the lattices in the population.

Atom permutation operator swaps chemical identities of atoms in randomly selected pairs, see Figure 7.7, while lattice remains unchanged. Such swaps provide the algorithm with steps in the search space that are very far and difficult in Euclidean distance. This operator is especially useful for systems, where chemically similar atoms are present.

Heredity operators are vital part of any evolutionary algorithm. If only mutation operators are present, then the algorithm is identical to a sophisticated version of a random search (such as D. Wales's basin hopping method [18]). Heredity operators are responsible for utilizing and refining the information about the system that we gather during the execution of the algorithm. Since properties of the crystal are determined by spatial arrangement of atoms in the unit cell, the most physically meaningful way to build a heredity operator is to conserve the information from parents by using spatially coherent pieces (spatial heredity).

To create a child from two parents, the algorithm first randomly chooses the lattice vector and a point on that vector. Then the unit cells of the parent structures are cut by the plane parallel to other vectors that goes through this point. Planar slices are matched, see Figure 7.8, and the number of atoms of each kind is adjusted. For big cells, it is possible to use more than two structures as parents and combine slices from all of them into a single child structure. The lattice of the child structure is a weighted averaged of parent lattices.

Altogether three variation operators described above explore the search space while preserving and refining the good spatial features through generations. It can be visualized by comparing the best structures from different generations, see, for example, Figures 7.9 and 7.10.

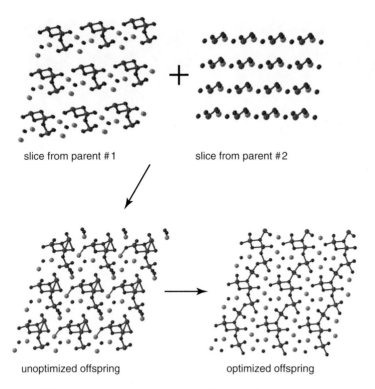

Figure 7.8 Heredity operator combines spatial slices from different parent structures to form an offspring structure. From Oganov et al. [9].

7.1.6
Survival of the Fittest and Selection of Parents

When the offspring structures are produced, we have to create a new generation. Usually it is a good idea to select the best few structures from the previous generation into the new one. This will make sure that we would not lose the best structures found during the algorithm execution and is also helpful if we want to search for low-energy metastable states, in addition to the global minimum. New generation is filled with the offspring structures, which are obtained from parent structures using variation operators. But which structures are to be chosen as parents? We rank the solutions by their free energy, worst 40% of structures are discarded (this threshold can be changed by user), and the probability for structure among best 60% to be chosen as a parent is proportional to its fitness rank. Stochastic selection is usually slower than deterministic one, when we choose only the best offspring to pass selection. However, it maintains a relatively high diversity of the population, which saves the simulation from being "trapped" in the basin of attraction of a good local minimum that is not the global one.

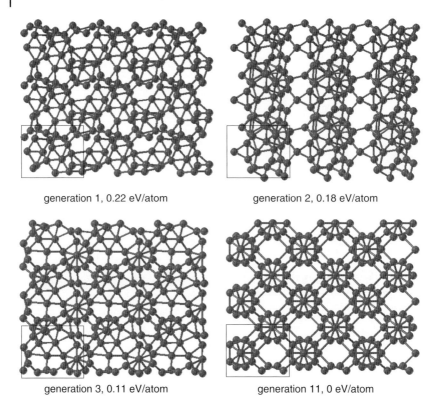

generation 1, 0.22 eV/atom

generation 2, 0.18 eV/atom

generation 3, 0.11 eV/atom

generation 11, 0 eV/atom

Figure 7.9 Best structures in generation from the evolutionary search for boron structure at 1 atm. The energy per atom shows the deviation from the ground state. One can see that parts of the icosahedra that build the most stable structure started to appear in the very first generations and were slowly refined during the run. From Oganov et al. [19].

7.1.7
Halting Criteria

After the new generation is produced and structures are relaxed, algorithm applies selection and variation operators to create the next one. This process is repeated until some halting criteria are achieved. Structure with the best enthalpy is considered as a candidate for the ground state of the system under given conditions. One can repeat the simulation a few times to increase the confidence of the results.

Obvious halting criterion would be reaching a certain number of subsequent generations where the best structure is not changing. However, sometimes we may need more sophisticated criteria. For example, we can continue calculations till the certain level of diversity is maintained and stop only when the population is filled with structures from a around the best structure. Alternatively, we may know something about the desired structure, for example, symmetry group or some other

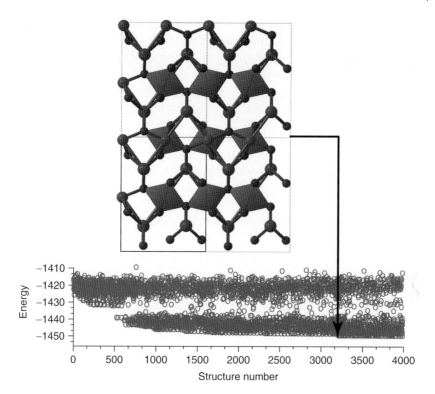

Figure 7.10 An example of a large test: prediction of the crystal structure of MgSiO$_3$ with 80 atoms in the supercell. The lower panel shows that the global-minimum structure was found within ∼3200 attempts. From Oganov and Glass [20].

information. In this case, we should continue calculation until the best structure has desired properties or the whole population is converged to a single funnel.

7.1.8
Premature Convergence and How to Prevent It: Fingerprint Function

As we already mentioned earlier, evolutionary algorithms always risk to be "trapped" in a funnel around some local minimum that is not a global optimum. This happens because good structures tend to produce children in their vicinity, filling the population with its own replicas and structures from their basin of attraction, reducing the diversity of the population. Such algorithm behavior (that we call "cancer growth phenomena") is especially common for energy landscapes with many good local minima surrounded by relatively low-energy areas that are separated from each other by high barriers. Sadly, this type of energy landscape is rather common. Therefore, a method that allows us to control the diversity of the population and avoid its premature convergence into one basin of attraction would increase the effectiveness of the algorithm.

The first question that arises when you try to develop such a method is "how can we detect similar structures and measure the degree of similarity between them?" Direct comparison of atomic coordinates will not work because they are represented in lattice vectors units and there are many equivalent ways to choose a unit cell. Free energy difference will also fail to describe the similarity between the structures because the search space is too noisy and not monotonic. We need some method that does not depend on unit cell choice and is determined solely by structure (and not representation). Ideally, at the same time, it should distinguish the structures where two atoms of different sorts are swapped in their positions. It should also be robust to small numerical errors that are unavoidable in real calculations. Nowadays scientists start to explore such methods. The most common approach to capture the essential geometrical properties of the structure is to use a specially constructed crystal structure descriptor, for example, radial distribution function (RDF).

In our algorithm, we use so-called fingerprint function [21, 22] to describe a crystal structure. It is a function, related to RDF and diffraction spectra, defined as

$$f(R) = \sum_i \sum_{j \neq i} \frac{Z_i Z_j}{4\pi R_{ij}^2} \frac{V}{N} \delta(R - R_{ij})$$

Here Z_i is the atomic number for atom i, R_{ij} is the distance between atoms i and j, V is the unit cell volume, and N is the number of atoms in the unit cell. We would like to note that R is a variable, not a parameter. The index i goes over all atoms in the unit cell, while index j goes over all atoms within some cutoff distance from the atom i. To remove fingerprint dependency from cutoff distance, the function is normalized as follows:

$$f_n(R) = \frac{f(R)}{\sum_{i,j} Z_i Z_j N_i N_j} - 1$$

Here N_i is the number of atoms in the unit cell with atomic number Z_i and the two sums go over all distinct Z values.

Fingerprint function as a method to describe the crystal structure has all the desired properties listed above. First of all, it does not depend on absolute atomic coordinates, but only on interatomic distances. Therefore, the choice of unit cell will not influence $f(R)$. Small perturbations of atomic positions will influence fingerprint function only slightly. Using atomic numbers as weighting coefficients allows us to take into account atom ordering. One could also measure the similarity between structures by computing the distance between their fingerprint functions. In our algorithm we used cosine distance, but one could use other metrics as well, for example, Cartesian distance or Minkowski norm [21].

To simplify and speed up the calculations, we discretize the fingerprint function and represent it as a vector *FP*, called fingerprint.

$$FP_i(R) = \frac{1}{D} \int_{iD}^{(i+1)D} f_n(R) dR$$

Cosine distance between two fingerprints for structures *i* and *j* is then defined as

$$d_{ij} = 0.5 \left(1 - \frac{FP_i FP_j}{\|FP_i\| \|FP_j\|}\right)$$

Having provided the fingerprint's space with a distance measure, we could group fingerprints, and thus structures, using a "similarity" or "almost equality" criteria.

7.1.9
Improved Selection Rules and Heredity Operator

The ability to measure the degree of similarity between structures allows us to improve the selection rules and variation operators described above [15]. First of all, determining the structures that will participate in building the next generation we ignore all similar and choose only different ones. Two structures are considered "similar" if the distance between their fingerprints is less than some user-defined threshold. This will increase the diversity of the offspring population and preclude premature convergence into a single funnel.

We already mentioned that letting a few best structures to survive and become members of the next generation is helpful if we search for metastable states and other low-energy (but not ground state) solutions. This can also be extremely useful for improving global optimization, as it increases the learning power of the algorithm. However, to avoid trapping in the local minimum we have to limit the number of surviving structures per basin of attraction. In most cases, we do not want more than a few or even just one structure per funnel. Ability to measure the similarity between structures allows us to set an additional "similarity" threshold and not letting structures with distance less than this threshold survive for the next population.

Heredity operator can also be improved using fingerprints to reduce the number of poor offsprings. In most cases if we take two good parents from different funnels, their offspring will be extremely bad. This phenomenon is schematically explained in Figure 7.11. This is especially useful for calculations where we let many good structures from different funnels survive, thus increasing the probability of choosing too different parents. The simplest way to avoid this is to introduce a threshold for maximum distance between parents that are allowed to produce a child. This trick is similar to "niching," [23] used in cluster structure prediction calculations (though we use more universal and nonempirical way of niching). For clusters, niching proved to be a major improvement.

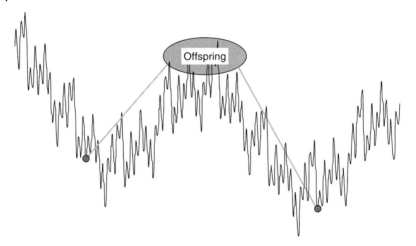

Figure 7.11 Two parents from different basins of attraction that have quite different structures usually produce a poor offspring in the high-energy areas between the funnels.

7.1.10
Extension to Molecular Crystals

In case of many organic crystals, assembling the most stable structure from single atoms will give ... a mixture of H_2O, CO_2, and perhaps other simple molecules, for a simple reason that most complex organic substances are metastable rather than thermodynamically stable. Those complex organic structures are possibly thermodynamically stable only under constraint of fixed (or partially fixed, with some conformational freedom) molecules as building blocks.

Unlike atoms, molecules are no longer point particles with spherical symmetry. Molecular handling capabilities have been implemented in USPEX [24], and are a straightforward extension of the atomic case. For variation operators molecules are now represented by a center of mass and orientation angles relative to a Cartesian coordinate frame. Offspring structures partially inherit the molecular orientations and center-of-mass positions. In addition to normal variation operators, we apply rotational mutation (whereby a randomly selected molecule within the unit cell is rotated as a whole along one or more axes). Special case must now be taken to avoid molecular overlap that may destroy the molecules upon relaxation. Plus, at least during the first stages of local optimization, the molecules should be kept fixed.

7.1.11
Adaptation to Clusters

Schönborn *et al.* [25] have translated the original version of the method [8, 26] to cluster structure prediction. The resulting algorithm was probably more effective than the earlier pioneering work by Deaven and Ho [27], and outperformed the minima hopping method in its original formulation [28]. However, with some new developments the latter algorithm was able to show similar, and in some cases superior, performance to our evolutionary algorithm. It remains to be seen how the relative performance will change if additional ingredients (fingerprints, niching) are incorporated in the cluster prediction method. Our first results in this direction are extremely encouraging. Adaptation to Clusters.

7.1.12
Extension to Variable Compositions: Toward Simultaneous Prediction of Stoichiometry and Structure

Another major extension of the method is to enable simultaneous prediction of all stable stoichiometries and structures (in a given range of compositions). We would like to mention the pioneering study by Jóhannesson et al. [29], who succeeded in predicting stable stoichiometries of alloys within a given structure type. For the completely unconstrained search for both the stoichiometry and structure, a preliminary method outline was proposed in Ref. [30] and implemented in Refs. [31, 32]. Here, the basic ideas are as follows:

1) Start with a population randomly (and sparsely) sampling the whole range of compositions of interest,
2) Allow variation operators to change chemical composition (we lift chemistry-preserving constraints in the heredity operator and, in addition to the permutation operator, introduce a "chemical transmutation" operator),
3) Evaluate the quality of each structure not by its (free) energy, but the (free) energy per atom minus the (free) energy of the most stable isochemical mixture of already sampled compounds. This means that this fitness function depends on history of the simulation.

Such an approach seems to work (Figure 7.12), but requires further major developments. While [31] introduced a constraint that in each simulation the total number of atoms in the unit cell is fixed, our method has no such constraint, and this proves beneficial and very convenient. An example of a (very difficult) system is given in Figure 7.13. Odd as it may seem, a binary Lennard–Jones system with a 1 : 2 ratio of radii (see caption to Figure 7.13 for details of the model) exhibits a large number of ground states – including the exotic $A_{14}B$ compound and the well-known AlB_2-type structure, and several marginally unstable compositions (such as A_8B_7, $A_{12}B_{11}$, A_6B_7, A_3B_4, AB_2). The correctness of these predictions is illustrated by the fact that a fixed-composition simulation at AB_2 stoichiometry produced results (gray square in Figure 7.13a) perfectly consistent with the variable-composition runs.

Figure 7.14 shows preliminary results for the Fe–Mg system at pressures of the Earth's inner core. In agreement with a recent work ([33], who arrived at this conclusion using different methods), we find that addition of Mg stabilizes the bcc structure and many of the intermediate compositions are bcc-based alloys, even though pure Fe has an hcp ground state at this pressure.

7 Crystal Structure Prediction Using Evolutionary Approach

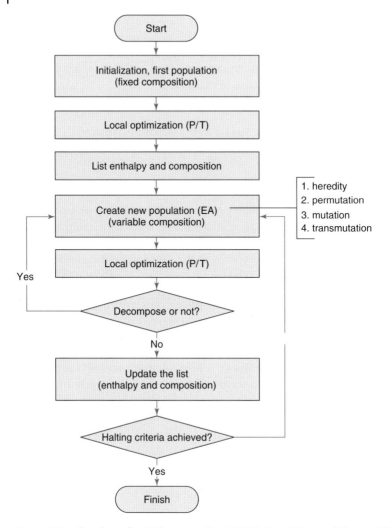

Figure 7.12 Flowchart of variable-composition USPEX. From Wang and Oganov [30].

7.2
A Few Illustrations of the Method

Any method is worth as much as its applications are, and here we review only a small selection of results obtained with USPEX, with the purpose of demonstrating its utility and with the desire to mention some interesting physics uncovered by it. All calculations described here were performed within the generalized gradient approximation (GGA; [34]) and the PAW method [35, 36], using the VASP code [12] for local optimization and total energy calculations. The predicted structures correspond to the global minima on the approximate free energy landscapes – that is, true ground states in cases where the GGA gives an adequate description of the system.

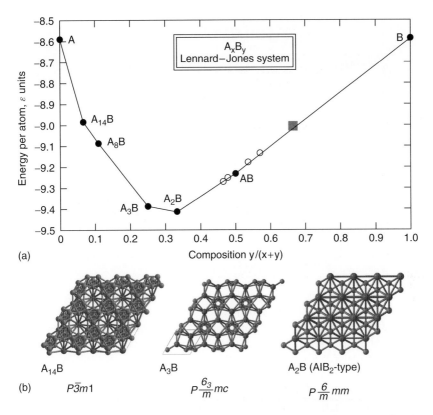

Figure 7.13 Variable-composition USPEX simulation of the A_xB_y binary Lennard–Jones system. (a) Filled circles – stable compositions, open circles – marginally unstable compositions (A_8B_7, $A_{12}B_{11}$, A_6B_7, A_3B_4). Gray square – fixed-composition result for AB_2 stoichiometry, finding a marginally unstable composition in agreement with the variable-composition results. (b) Some of the stable structures. While the ground state of the one-component Lennard–Jones crystal is hexagonal close packed (hcp) structure, ground states of the binary Lennard–Jones system are rather complex (e.g., $A_{14}B$). Note that the structure found for A_2B belongs to the well-known AlB_2 structure type. The potential is of the Lennard–Jones form for each atomic ij-pair:

$$U_{ij} = \varepsilon_{ij}\left[\left(\frac{R_{min,ij}}{R}\right)^{12} - 2\left(\frac{R_{min,ij}}{R}\right)^{6}\right],$$

where $R_{min,ij}$ is the distance at which the potential reaches minimum, and ε is the depth of the minimum. In these simulations, we use additive atomic dimensions: $R_{min,BB} = 1.5 R_{min,AB} = 2 R_{min,AA}$ and nonadditive energies (to favor compound formation): $\varepsilon_{AB} = 1.25\varepsilon_{AA} = 1.25\varepsilon_{BB}$. From Oganov et al. [72].

7.2.1
Elements

7.2.1.1 Boron: Novel Phase with a Partially Ionic Character

Boron is perhaps the most enigmatic element: at least 16 phases were reported in the literature, but most are believed or suspected to be compounds (rather than forms of the pure element), and until recently the phase diagram was unknown.

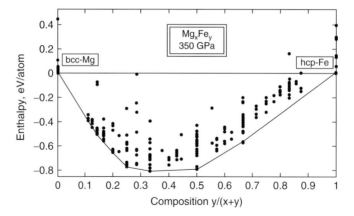

Figure 7.14 Mg–Fe system at 350 GPa. Circles indicate structures (their compositions and enthalpies) sampled, relative to the isochemical mixture of hcp-Fe and bcc-Mg. Clearly, Fe and Mg form very stable alloys (many of which have bcc-type structures) at this pressure. This variable-composition calculation was performed at the GGA level of theory [34].

Following experimental findings of J. Chen and V.L. Solozhenko (both arrived independently at the same conclusions in 2004) of a new phase at pressures above 10 GPa and temperatures of 1800–2400 K, the structure could not be determined from experimental data alone. We found the structure by using USPEX. We named this phase γ-B_{28} (because it contains 28 atoms/cell). Its structure (Figure 7.15) consists of icosahedral B_{12} clusters and B_2 pairs in an NaCl-type arrangement and exhibits sizable charge transfer from B_2 pairs to B_{12} clusters (Figure 7.16), quite unexpected for a pure element (see [19], for details).

γ-B_{28} can be represented as a "boron boride" $(B_2)^{\delta+}(B_{12})^{\delta-}$. While the exact value of the charge transfer δ depends on the definition of an atomic charge, all

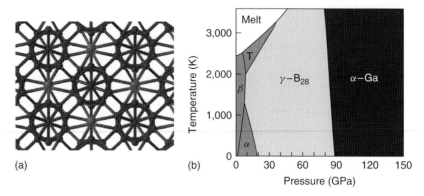

Figure 7.15 Boron: (a) structure of γ-B_{28} (B_{12} icosahedra and B_2 pairs are marked by different shades). (b) Phase diagram of boron, showing a wide stability field of γ-B_{28}. From Oganov et al. [19].

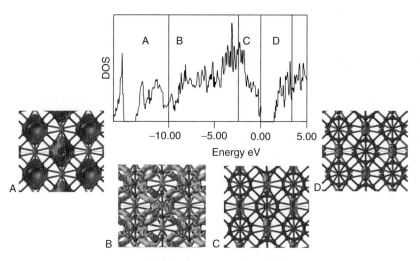

Figure 7.16 γ-B_{28}: total electronic DOS and energy-decomposed electron densities. Lowest-energy valence electrons are dominated by the B_{12} icosahedra, while top of the valence band and bottom of the conduction band (i.e., holes) are localized on the B_2 pairs. This is consistent with atom-projected DOSs and the idea of charge transfer $B_2 \rightarrow B_{12}$. From Oganov *et al.* [73].

definitions give the same qualitative picture. Our preferred definition, due to Bader [37], gives $\delta \sim 0.5$ [19]. Based on the similarity of synthesis conditions and many diffraction peaks, the same high-pressure boron phase may have been observed by Wentorf [38], though Wentorf's material was generally not believed to be pure boron (due to the sensitivity of boron to impurities and lack of chemical analysis or structure determination in his work) and its diffraction pattern was deleted from Powder Diffraction File database. γ-B_{28} is structurally related to several compounds – for instance, B_6P or $B_{13}C_2$, where the two sublattices are occupied by different chemical species (instead of interstitial B_2 pairs there are P atoms or C–B–C groups, respectively). Significant charge transfer can be found in other elemental solids, and observations of dielectric dispersion [39], implying LO–TO splitting, suggest it for β-B_{106}. The nature of the effect is possibly similar to γ-B_{28}. Detailed microscopic understanding of charge transfer in β-B_{106} would require detailed knowledge of its structure, and reliable structural models of β-B_{106} finally begin to emerge from computational studies [40–42]. γ-B_{28} is a superhard phase, with a measured Vickers hardness of 50 GPa [43], which puts it among the half a dozen hardest materials known to date.

7.2.1.2 Sodium: A Metal that Goes Transparent under Pressure

Sodium, a simple s-element at normal conditions, behaves in highly nontrivial ways under pressure. The discovery of an incommensurate host–guest structure [44], followed by the finding of several other complex phases [45] just below the pronounced minimum of sodium's melting curve, and the very existence of that extremely deep minimum in the melting curve at about 110 GPa [46] – all points to some unusual changes in the physics of sodium. Later, it was also shown that the incommensurate host–guest structure is a 1D metal [47], where conductivity is mainly within the guest atom chains.

Yet another unusual phenomenon was predicted using USPEX and later (but within the same paper – [48]) verified experimentally: on further compression sodium becomes a wide-gap insulator! This happens at ~190 GPa, and Figure 7.17 shows the crystal structure of the insulating "hP4" phase, its enthalpy relative to other structures, and the electronic structure. The structure contains two inequivalent Na positions, both six-coordinated: Na1 and Na2, which have the octahedral and trigonal-prismatic coordination, and the hP4 structure can be described as the elemental analog of the NiAs structure type. Calculations suggest that sodium is no longer an s-element; instead, the outermost valence electron has significant s-, p-, and d-characters (Figure 7.17c). In other words, sodium can be considered as a transition metal because of its significant d-character.

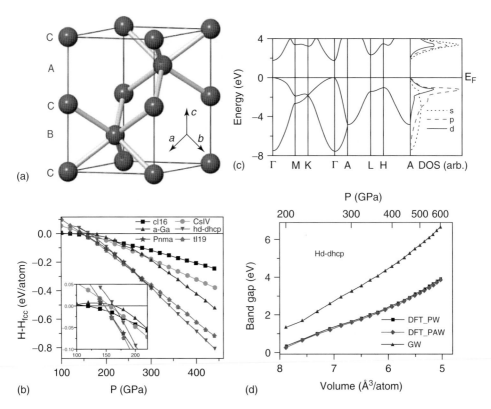

Figure 7.17 Transparent hP4 phase of sodium: (a) its crystal structure, (b) enthalpies of competing high-pressure phases (relative to the fcc structure), (c) band structure, and (d) pressure dependence of the band gap, indicating rapid increase in the band gap on compression. From Ma et al. [48].

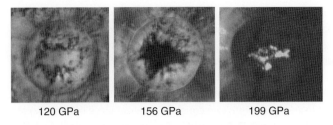

| 120 GPa | 156 GPa | 199 GPa |

Figure 7.18 Photographs of sodium samples under pressure. At 120 GPa, the sample is metallic and highly reflective, at 156 GPa the reflectivity is very low, and at 199 GPa the sample is transparent. From Ma et al. [48]. (Please find a color version of this figure on the color plates.)

The band gap is direct and increases with pressure. At 200 GPa, the band gap calculated with the GW approximation is 1.3 eV, and increases to 6.5 eV at 600 GPa. These predictions implied that above 200 GPa, sodium will be red and transparent, and at ~300 GPa it will become colorless and transparent! This has indeed been confirmed in experiment [48], see Figure 7.18. The insulating behavior is explained by the extreme localization of the valence electrons in the interstices of the structure, that is, the "empty" space (Figure 7.19). These areas of localization are characterized by surprisingly high values of the electron localization function (nearly 1.0) and maxima of the total electron density. The number of such maxima is half the number of sodium atoms, and therefore in a simple model we can consider Na atoms as completely ionized (Na^+), and interstitial maxima as containing one electron pair. The hP4 structure can also be described as a Ni_2In-type structure, where Na atoms occupy positions of Ni atoms, and interstitial electron pairs in

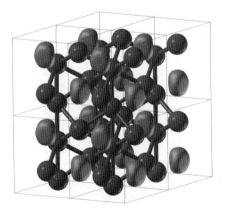

Figure 7.19 Crystal structure and electron localization function (isosurface contour 0.90) of the hP4 phase of sodium at 400 GPa. Interstitial electron localization is clearly seen. (Please find a color version of this figure on the color plates.)

hP4-Na sit on the same positions as In atoms in Ni_2In. At first counterintuitively, the degree of localization of the interstitial electron pairs increases with pressure, explaining the increase in the band gap (Figure 7.17d). hP4-Na can be described as an electride, that is, an ionic "compound" formed by ionic cores and localized interstitial electron pairs. The very fact that sodium, one of the best and simplest metals, under pressure becomes a transparent insulator with localized valence electrons is remarkable and forces one to reconsider classical ideas of chemistry.

Interstitial charge localization can be described in terms of (s)-p-d orbital hybridizations, and its origins are in the exclusionary effect of the ionic cores on valence electrons: valence electrons, feeling repulsion from the core electrons, are forced into the interstitial regions at pressures where atomic cores begin to overlap [50].

7.2.1.3 Superconducting ξ-Oxygen

Oxygen shows many unusual features under pressure. The metallic (superconducting at very low temperatures, [51]) ζ-phase, stable above 96 GPa, was discovered in 1995 [52], and its structure remained controversial for a long time. Neutron diffraction showed [53] that already in the ε-phase (at 8 GPa) there is no long-range magnetic order and likely even no local moments. The disappearance of magnetism is a consequence of increasing overlap of molecular orbitals with increasing pressure. Ultimately, orbital overlap leads to metallization.

Evolutionary simulations at 130 and 250 GPa uncovered two interesting structures with $C2/m$ and $C2/c$ space groups [54]. These have very similar enthalpies; the $C2/m$ structure is slightly lower in enthalpy and matches experimental X-ray diffraction and Raman spectroscopy data very well, better than the $C2/c$ structure [54]. Recently, a single-crystal X-ray diffraction study [55] confirmed our predicted $C2/m$ structure of ζ-oxygen. This structure is isosymmetric with the lower pressure ε-phase; both structures are shown in Figure 7.20.

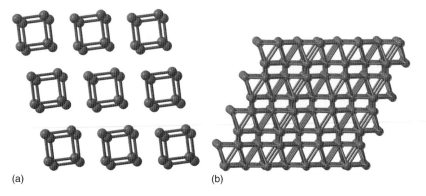

Figure 7.20 High-pressure structures of oxygen: (a) experimentally found ε-O_8 structure at 17.5 GPa [56]. (b) Predicted and confirmed $C2/m$ structure of the ζ-phase at 130 GPa [54]. Contacts up to 2.2 Å are shown as bonds.

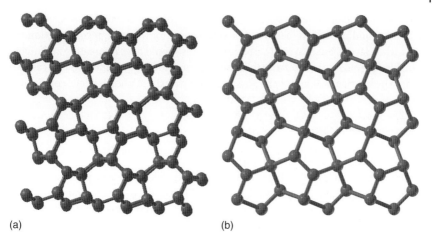

Figure 7.21 Structures of carbon: (a) novel metastable form with a 3D-framework and (b) hypothetical 2D structure, possible at interfaces. From Oganov et al. [9] and Oganov and Glass [26].

7.2.1.4 Briefly on Some of the (Many) Interesting Carbon Structures

While thermodynamically stable phases of carbon are few and simple (but still far from boring!), the metastable ones are almost inexhaustible and worth exploring both computationally and experimentally. USPEX is an attractive method for predicting low-energy metastable structures (i.e., not only the global minimum, but also the nearby local ones) because of the construction of the method, focusing to an increasing degree on low-energy parts of the energy landscape as the simulation progresses.

A number of interesting metastable carbon structures derived by USPEX was mentioned in Ref. [26], and here we just wish to mention a few novel twists in the story. One of the low-energy structures found in Ref. [26] was the three-dimensional "5 + 7" structure shown in Figure 7.21a. Later, it was found [57] that this structure matches the observed properties of the so-called superhard graphite, a new phase of carbon obtained by cold compression of graphite to pressures above 15 GPa, that is, beyond the stability field of graphite [58]. The "5 + 7" structure contains corrugated hexagonal layers reminiscent of the graphene layers; its predicted density and hardness are similar to diamond's, and its diffraction pattern closely agrees with experimental one for "superhard graphite" of Ref. [58].

Another set of interesting carbon structures was found [9] in 2D-space (hence, such structures might appear on surfaces and in interfaces) – looking for possible stable compounds in the Xe–C system at high pressures, we consistently found layered structures with Xe and C atoms segregated from each other, indicative of the tendency to phase separation. While the Xe layers had the expected close-packed configuration, the layers of carbon atoms had very creative arrangements made of three- and four-connected carbon atoms. One of such layers is shown in Figure 7.21b. While no stable Xe–C compounds were found at pressures of up to

200 GPa, we did observe in the segregated structures a considerable Xe–C bonding. It is this bonding that we believe to be responsible for the large reconstructions in the carbon layers. Such layered carbon structures should have interesting and highly tunable (by means of modifying their interaction with the substrate, in this case Xe layers) properties, and it must be possible to prepare them under special conditions.

7.2.2
Compounds and Minerals

7.2.2.1 Insulators by Metal Alloying?

Based on the electronic "jellium" model, the existence of a very interesting class of *insulating* materials (e.g., $Al_{13}K$ and $Al_{12}C$), based on icosahedral aluminum clusters, has been hypothesized (e.g., in Ref. [59]). Recent work done with USPEX [20] has refuted this suggestion. For instance, for $Al_{12}C$ a much more stable structure exists – fcc structure with stacking faults, where carbon atoms occupy octahedrally coordinated sites (Figure 7.22). This also suggests that addition of carbon to aluminum may promote the formation of stacking faults, a conclusion of some interest to metallurgy.

7.2.2.2 MgB$_2$: Analogy with Carbon and Loss of Superconductivity under Pressure

MgB$_2$ is a superconductor with a very high $T_c = 39$ K [60], highest among the experimentally studied conventional superconductors. In its structure, Mg atoms are sandwiched between graphene-like layers of boron atoms. Using USPEX, [49] found that at high pressure (190 GPa) MgB$_2$ adopts another structure (Figure 7.23) with B atoms forming a distorted hexagonal diamond structure. In both phases, holes are the main charge carriers, but the high-pressure phase is a much poorer

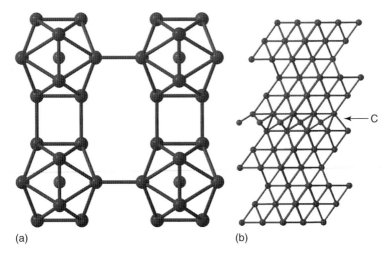

Figure 7.22 Structures of $Al_{12}C$ with 13 atoms/cell: (a) with icosahedral clusters and (b) more stable stacking-fault structure. From Oganov and Glass [20].

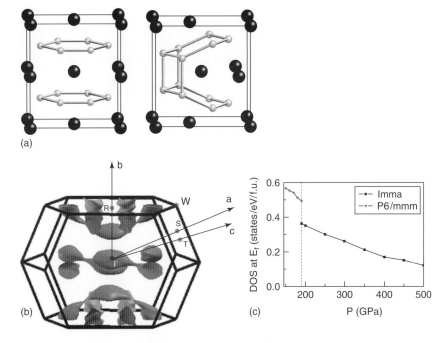

Figure 7.23 New phase of MgB$_2$: (a) its structure (right) and relationship to the structure known at 1 atm (left), (b) its Fermi surface, and (c) electronic density of states at the Fermi level as a function of pressure. From Ma et al. [49]. (Please find a color version of this figure on the color plates.)

metal and is not a superconductor. The structural trend parallels that of carbon: graphite is a much better conductor than diamond.

7.2.2.3 Hydrogen-Rich Hydrides under Pressure, and Their Superconductivity

Recently, three studies were published, where USPEX was used to find high-pressure metallic ground states of hydrogen-rich compounds – GeH$_4$ [61], SiH$_4$ [62], and LiH$_2$, LiH$_6$, and LiH$_8$ [63]. In line with the recent proposals of N.W. Ashcroft, we indeed found that hydrides adopt superconducting states more readily than pure hydrogen does.

Already at 100 GPa, LiH$_n$ ($n > 1$) metallic compounds attain thermodynamic stability. Their structures (Figure 7.24) contain the "semimolecular" H$_2$ units with bonds slightly longer than 0.74 Å of the free H$_2$ molecule. Zurek et al. [63] have explained that it is charge transfer from Li to the unoccupied (antibonding) levels of the H$_2$ molecules that simultaneously creates the metallic character and weakens bonding within the H$_2$ units. We also note that the very existence (and stability!) of such compounds violate the traditional chemical valency concepts. LiH$_2$, by the way, has a structure closely related to the host–guest structures known for Ca, Sr, and Ba (H$_2$ units occupy the guest sublattice, with Li + H being the host).

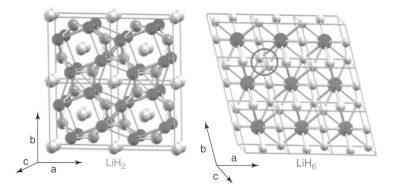

Figure 7.24 Structures of LiH_2 (a) and LiH_6 (b) predicted to be stable at pressures >100 GPa Li atoms are green, "lone" hydrogen atoms are pink, and those in the H_2 units are white. (from [63]). (Please find a color version of this figure on the color plates.)

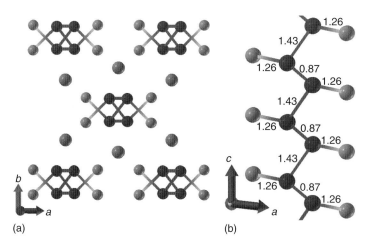

Figure 7.25 Crystal structure of the $C2/c$ phase of GeH_4. (a) Two types of hydrogen atoms are shown by different color (light – "lone" atoms, and dark – forming H_2 semi-molecular units). Panel (b) shows the connectivity of hydrogens in the structure and distances (in Å). (Please find a color version of this figure on the color plates.)

An interesting prediction was made for GeH_4 that it will become thermodynamically stable above 196 GPa [61] and that the stable $C2/c$ should have a very high superconducting T_c (64 K at the pressure of 220 GPa). The structure of this phase (Figure 7.25a) contains a branched chain of H atoms with alternating bond lengths shown in Figure 7.25b.

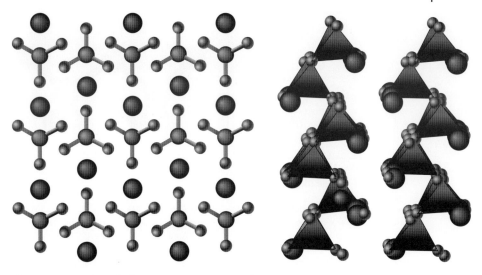

Figure 7.26 CaCO$_3$ at high pressure. (a) Structure of postaragonite phase and (b) C222$_1$ phase. From [65].

7.2.2.4 High-Pressure Polymorphs of CaCO$_3$

High-pressure behavior of carbonates is very important for understanding the Earth's carbon cycle, but remained controversial until recently. Even for CO$_2$, a chemically simpler compound, major controversies still exist (see [20], for a discussion of results on CO$_2$). For CaCO$_3$, there is a well-known transition from calcite to aragonite at ∼2 GPa, followed by a transition to a postaragonite phase at ∼40 GPa [64], the structure of which (Figure 7.26a) was solved [65] using USPEX, and the predicted structure matched the experimental X-ray diffraction pattern well. Now postaragonite structure is also known in SrCO$_3$ and BaCO$_3$ [64, 66].

Furthermore, we have predicted [65] that above 137 GPa a new phase, with space group C222$_1$ and containing chains of carbonate tetrahedral (Figure 7.26b), becomes stable. Recently this prediction was verified by experiments [67] at pressures above 130 GPa. We note that both postaragonite and the C222$_1$ structure (Figure 7.26b) belong to new structure types and could not have been found by analogy with any known structures.

The presence of tetrahedral carbonate ions at very high pressures invites an analogy with silicates, but the analogy is limited. In silicates, the intertetrahedral angle Si–O–Si is extremely flexible [68], which is one of the reasons for the enormous diversity of silicate structure types. Figure 7.27 shows the variation of the energy as a function of the Si–O–Si angle in the model H$_6$Si$_2$O$_7$ molecule – method borrowed from [68]. One can see only a shallow minimum at $\angle(Si-O-Si) = 135°$, but a deep minimum at $\angle(C-O-C) = 124°$ with steep energy variations for H$_6$C$_2$O$_6$ (Figure 7.27). This suggests a much more limited structural variety/flexibility of metacarbonates, compared to silicates. In tetrahedrally coordinated forms of both CaCO$_3$ (C222$_1$ structure) and CO$_2$ ($\beta-$ cristobalite structure, found to be stable

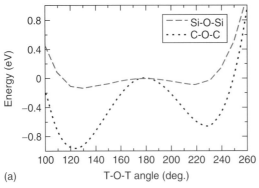

Figure 7.27 (a) Energy variation as a function of the T–O–T angle (red dashed line – T = Si, black dotted line – T = C). Calculations were performed on $H_6T_2O_7$ molecules; at each angle all T–O distances and O–T–O valence angles were optimized. Optimum angle C–O–C = 124°, Si–O–Si = 135°. (b) These calculations were performed with SIESTA code [13] using the GGA functional [34], norm-conserving pseudopotentials and a double-ζ basis set with a single polarization function for each atom. From [69].

above 19 GPa – [69–71]), the $\angle(C-O-C)$ angles are close to 124° in a wide pressure range.

7.3
Conclusions

Evolutionary algorithms, based on physically motivated forms of variation operators and local optimization, are a powerful tool enabling reliable and efficient prediction of stable crystal structures. This method has a wide field of applications in computational materials design (where experiments are time-consuming and expensive) and in studies of matter at extreme conditions (where experiments are very difficult or sometimes beyond the limits of feasibility).

One of the current limitations is the accuracy of today's *ab initio* simulations; this is particularly critical for strongly correlated and for van der Waals systems. Note, however, that the method itself does not make any assumptions about the way energies are calculated and can be used in conjunction with any method, that is able to provide total energies. Most of practical calculations are done at $T = 0$ K, but temperature can be included as long as the free energy can be calculated efficiently. Difficult cases are aperiodic and disordered systems (for which only the lowest energy periodic approximants and ordered structures can be predicted at this moment).

We are suggesting USPEX as the method of choice for crystal structure prediction of systems with up to $\sim$50 atoms/cell, where no information (or just the lattice parameters) is available. Above 50–100 atoms/cell runs become expensive due to the "curse of dimensionality" (although still feasible), eventually necessitating the use of other ideas within USPEX or another approach. There is, however, hope of enabling structure prediction for a very large (>200 atoms/cell) systems.

USPEX has been applied to many important problems. One expects many more applications to follow, both in high-pressure research and in materials design.

Acknowledgments

ARO and AOL thank Intel Corporation, DARPA (grant 54751), and Research Foundation of Stony Brook University for financial support. We thank the Joint Supercomputer Center (Russian Academy of Sciences, Moscow) for providing supercomputer time. USPEX code is available on request from ARO.

References

1. Gavezzotti, A. (1994) Are crystal structure predictable? *Acc. Chem. Res.*, **27**, 309–314.
2. Maddox, J. (1988) Crystals from first principles. *Nature*, **335**, 201.
3. Tsai, H.K., Yang, J.M., Tsai, Y.F., and Kao, C.Y. (2004) An evolutionary algorithm for large traveling salesman problems. *IEEE Trans. Syst. Man. Cybern. B Cybern.*, **34** (4), 1718–1729.
4. Bush, T.S., Catlow, C.R.A., and Battle, P.D. (1995) Evolutionary programming techniques for predicting inorganic crystal structures. *J. Mater. Chem.*, **5**, 1269–1272.
5. Woodley, S.M., Battle, P.D., Gale, J.D., and Catlow, C.R.A. (1999) The prediction of inorganic crystal structures using a genetic algorithm and energy minimization. *Phys. Chem. Chem. Phys.*, **1**, 2535–2542.
6. Woodley, S.M. (2004) Prediction of crystal structures using evolutionary algorithms and related techniques. *Struct. Bond.*, **110**, 95–132.
7. Bazterra, V.E., Ferraro, M.B., and Facelli, J.C. (2002) Modified genetic algorithm to model crystal structures. I. Benzene, naphthalene and anthracene. *J. Chem. Phys.*, **116**, 5984–5991.
8. Glass, C.W., Oganov, A.R., and Hansen, N. (2006) USPEX – evolutionary crystal structure prediction. *Comput. Phys. Commun.*, **175**, 713–720.
9. Oganov, A.R., Ma, Y., Glass, C.W., and Valle, M. (2007) Evolutionary crystal structure prediction: overview of the USPEX method and some of its applications. *Psi-k Newsl.*, **84**, 142–171.
10. Valle, M. (2005) STM3: a chemistry visualization platform. *Z. Krist.*, **220**, 585–588.
11. Valle, M. and Oganov, A.R. (2010) Crystal fingerprints space, A novel paradigm to study crystal structures sets (in press). *Acta Cryst. A*.
12. Kresse, G. and Furthmüller, J. (1996) Efficient iterative schemes for *ab initio* total-energy calculations using a plane wave basis set. *Phys. Rev. B*, **54**, 11169–11186.
13. Soler, J.M., Artacho, E., Gale, J.D., Garcia, A., Junquera, J., Ordejon, P., and Sanchez-Portal, D. (2002) The SIESTA method for *ab initio* order-N materials simulation. *J. Phys.: Condens. Matter*, **14**, 2745–2779.
14. Gale, J.D. (2005) GULP: capabilities and prospects. *Z. Krist.*, **220**, 552–554.
15. Lyakhov, A.O., Oganov, A.R., and Valle, M. (2010) How to predict very large and complex crystal structures. *Comput. Phys. Commun.*, **181**, 1623–1632.
16. Lloyd, L.D. and Johnston, R.L. (1998) Modelling aluminum clusters with an empirical many-body potential. *Chem. Phys.*, **236**, 107.

17. Pickard, C.J. and Needs, R.J. (2006) High-pressure phases of silane. *Phys. Rev. Lett.*, **97**, 045504.
18. Wales, D.J. and Doye, J.P.K. (1997) Global optimization by Basin–Hopping and the lowest energy structures of Lennard–Jones clusters containing up to 110 atoms. *J. Phys. Chem. A*, **101**, 5111.
19. Oganov, A.R., Chen, J., Gatti, C., Ma, Y.-Z., Ma, Y.-M., Glass, C.W., Liu, Z., Yu, T., Kurakevych, O.O., and Solozhenko, V.L. (2009) Ionic high-pressure form of elemental boron. *Nature*, **457**, 863–867.
20. Oganov, A.R. and Glass, C.W. (2008) Evolutionary crystal structure prediction as a tool in materials design. *J. Phys.: Cond. Mattter*, **20**, 064210.
21. Valle, M. and Oganov, A.R. (2008a) Crystal structure classifier for an evolutionary algorithm structure predictor. Proceedings of the IEEE Symposium on Visual Analytics Science and Technology, October 21–23, 2008, Columbus, pp. 11–18.
22. Oganov, A.R. and Valle, M. (2009) How to quantify energy landscapes of solids. *J. Chem. Phys.*, **130**, 104504.
23. Hartke, B. (1999) Global cluster geometry optimization by a phenotype algorithm with Niches: location of elusive minima, and low-order scaling with cluster size. *J. Comput. Chem.*, **20**, 1752.
24. Glass, C.W. (2008) Computational crystal structure prediction. PhD thesis, ETH Zurich.
25. Schönborn, S., Goedecker, S., Roy, S., and Oganov, A.R. (2009) The performance of minima hopping and evolutionary algorithms for cluster structure prediction. *J. Chem. Phys.*, **130**, 144108.
26. Oganov, A.R. and Glass, C.W. (2006) Crystal structure prediction using *ab initio* evolutionary techniques: principles and applications. *J. Chem. Phys.*, **124**, 244704.
27. Deaven, D.M. and Ho, K.M. (1995) Molecular geometry optimization with a genetic algorithm. *Phys. Rev. Lett.*, **75**, 288–291.
28. Gödecker, S. (2004) Minima hopping: An efficient search method for the global minimum of the potential energy surface of complex molecular systems. *J. Chem. Phys.*, **120**, 9911–9917.
29. Jóhannesson, G.H., Bligaard, T., Ruban, A.V., Skriver, H.L., Jacobsen, K.W., and Nørskov, J.K. (2002) Combined electronic structure and evolutionary search approach to materials design. *Phys. Rev. Lett.*, **88**, 255506.
30. Wang, Y. and Oganov, A.R. (2008) Research on the evolutionary prediction of very complex crystal structures. IEEE Computational Intelligence Society Walter Karplus. Summer Research Grant, Final Report. (ieee-cis.org/_files/EAC_Research_2008_Report_WangYanchao.October 2008).
31. Trimarchi, G., Freeman, A.J., and Zunger, A. (2009) Predicting stable stoichiometries of compounds via evolutionary global space-group optimization. *Phys. Rev. B*, **80**, 092101.
32. Lyakhov, A.O., and Oganov, A.R. (2010) Simultaneous prediction of chemical formula and crystal structure using an evolutionary algorithm (in press).
33. Kadas, K., Vitos, L., Johansson, B., and Ahuja, R. (2009) Stability of body-centered cubic iron-magnesium alloys in the Earth's inner core. *Proc. Natl. Acad. Sci.*, **106**, 15560–15562.
34. Perdew, J.P., Burke, K., and Ernzerhof, M. (1996) Generalized gradient approximation made simple. *Phys. Rev. Lett.*, **77**, 3865–3868.
35. Blöchl, P.E. (1994) Projector augmented-wave method. *Phys. Rev. B*, **50**, 17953–17979.
36. Kresse, G. and Joubert, D. (1999) From ultrasoft pseudopotentials to the projector augmented-wave method. *Phys. Rev. B*, **59**, 1758–1775.
37. Bader, R. (1990) *Atoms in Molecule: A Quantum Theory*, Oxford University Press, Oxford.
38. Wentorf, R.H. (1965) Boron: another form. *Science*, **147**, 49–50.
39. Tsagareishvili, O.A., Chkhartishvili, L.S., and Gabunia, D.L. (2009) Apparent low-frequency charge capacitance of semiconducting boron. *Semiconductors*, **43**, 14–20.
40. van Setten, M.J., Uijttewaal, M.A., de Wijs, G.A., and de Groot, R.A. (2007) Thermodynamic stability of boron: The

role of defects and zero point motion. *J. Am. Chem. Soc.*, **129**, 2458–2465.

41. Widom, M. and Mikhalkovic, M. (2008) Symmetry-broken crystal structure of elemental boron at low temperature. *Phys. Rev. B*, **77**, 064113.

42. Ogitsu, T., Gygi, F., Reed, J., Motome, Y., Schwegler, E., and Galli, G. (2009) Imperfect crystal and unusual semiconductor: boron, a frustrated element. *J. Am. Chem. Soc.*, **131**, 1903–1909.

43. Solozhenko, V.L., Kurakevych, O.O., and Oganov, A.R. (2008) On the hardness of a new boron phase, orthorhombic $\gamma - B_{28}$. *J. Superhard Mater.*, **30**, 428–429.

44. Hanfland, M., Syassen, K., Loa, I., Christensen, N.E., and Novikov, D.L. (2002) Na at megabar pressures. Poster at 2002 High Pressure Gordon Conference.

45. Gregoryanz, E., Lundegaard, L.F., McMahon, M.I., Guillaume, C., Nelmes, R.J., and Mezouar, M. (2008) Structural diversity of sodium. *Science*, **320**, 1054–1057.

46. Gregoryanz, E., Degtyareva, O., Somayazulu, M., Hemley, R.J., and Mao, H.K. (2005) Melting of dense sodium. *Phys. Rev. Lett.*, **94**, 185502.

47. Lazicki, A., Goncharov, A.F., Struzhkin, V.V., Cohen, R.E., Liu, Z., Gregoryanz, E., Guillaume, C., Mao, H.K., and Hemley, R.J. (2009) Anomalous optical and electronic properties of dense sodium. *Proc. Natl. Acad. Sci.*, **106**, 6525–6528.

48. Ma, Y., Eremets, M.I., Oganov, A.R., Xie, Y., Trojan, I., Medvedev, S., Lyakhov, A.O., Valle, M., and Prakapenka, V. (2009a) Transparent dense sodium. *Nature*, **458**, 182–185.

49. Ma, Y., Wang, Y., and Oganov, A.R. (2009b) Absence of superconductivity in the novel high-pressure polymorph of MgB_2. *Phys. Rev. B*, **79**, 054101.

50. Neaton, J.B. and Ashcroft, N.W. (1999) Pairing in dense lithium. *Nature*, **400**, 141–144.

51. Shimizu, K., Suhara, K., Ikumo, M., Eremets, M.I., and Amaya, K. (1998) Superconductivity in oxygen. *Nature*, **393**, 767–769.

52. Akahama, Y., Kawamura, H., Hausermann, D., Hanfland, M., and Shimomura, O. (1995) New high-pressure structural transition of oxygen at 96 GPa associated with metalization in a molecular solid. *Phys. Rev. Lett.*, **74**, 4690–4693.

53. Goncharenko, I.N. (2005) Evidence for a magnetic collapse in the epsilon phase of solid oxygen. *Phys. Rev. Lett.*, **94**, 205701.

54. Ma, Y.-M., Oganov, A.R., and Glass, C.W. (2007) Structure of the metallic ζ-phase of oxygen and isosymmetric nature of the $\varepsilon - \zeta$ phase transition: Ab initio simulations. *Phys. Rev. B*, **76**, 064101.

55. Weck, G., Desgreniers, S., Loubeyre, P., and Mezouar, M. (2009) Single-crystal structural characterization of the metallic phase of oxygen. *Phys. Rev. Lett.*, **102**, 255503.

56. Lundegaard, L.F., Weck, G., McMahon, M.I., Desgreniers, S., and Loubeyre, P. (2006) Observation of an O_8 molecular lattice in the epsilon phase of solid oxygen. *Nature*, **443**, 201–204.

57. Li, Q., Oganov, A.R., Wang, H., Wang, H., Xu, Y., Cui, T., Ma, Y., Mao, H.-K., and Zou, G. (2009) Superhard monoclinic polymorph of carbon. *Phys. Rev. Lett.*, **102**, 175506.

58. Mao, W.L., Mao, H.K., Eng, P.J., Trainor, T.P., Newville, M., Kao, C.C., Heinz, D.L., Shu, J., Meng, Y., and Hemley, R.J. (2003) Bonding changes in compressed superhard graphite. *Science*, **302**, 425–427.

59. Gong, X.G. (1997) Structure and stability of cluster-assembled solid $Al_{12}C(Si)$: a first-principles study. *Phys. Rev. B*, **56**, 1091–1094.

60. Nagamatsu, J., Nakagawa, N., Muranaka, T., Zenitani, Y., and Akimitsu, J. (2001) Superconductivity at 39 K in magnesium diboride. *Nature*, **410**, 63–64.

61. Gao, G., Oganov, A.R., Bergara, A., Martinez-Canalez, M., Cui, T., Iitaka, T., Ma, Y., and Zou, G. (2008) Superconducting high pressure phase of germane. *Phys. Rev. Lett.*, **101**, 107002.

62. Martinez-Canales, M., Oganov, A.R., Lyakhov, A., Ma, Y., and Bergara, A.

(2009) Novel structures of silane under pressure. *Phys. Rev. Lett.*, **102**, 087005.
63. Zurek, E., Hoffmann, R., Ashcroft, N.W., Oganov, A.R., and Lyakhov, A.O. (2009) A little bit of lithium does a lot for hydrogen. *Proc. Natl. Acad. Sci.*, **106**, 17640–17643.
64. Ono, S., Kikegawa, T., Ohishi, Y., and Tsuchiya, J. (2005) Post-aragonite phase transformation in $CaCO_3$ at 40 GPa. *Am. Mineral.*, **90**, 667–671.
65. Oganov, A.R., Glass, C.W., and Ono, S. (2006) High-pressure phases of $CaCO_3$: Crystal structure prediction and experiment. *Earth Planet. Sci. Lett.*, **241**, 95–103.
66. Ono, S. (2007) New high-pressure phases in $BaCO_3$. *Phys. Chem. Miner.*, **34**, 215–221.
67. Ono, S., Kikegawa, T., and Ohishi, Y. (2007) High-pressure phase transition of $CaCO_3$. *Am. Mineral.*, **92**, 1246–1249.
68. Lasaga, A.C. and Gibbs, G.V. (1987) Applications of quantum-mechanical potential surfaces to mineral physics calculations. *Phys. Chem. Miner.*, **14**, 107–117.
69. Oganov, A.R., Ono, S., Ma, Y., Glass, C.W., and Garcia, A. (2008) Novel high-pressure structures of $MgCO_3$, $CaCO_3$ and CO_2 and their role in the Earth's lower mantle. *Earth Planet. Sci. Lett.*, **273**, 38–47.
70. Dong, J.J., Tomfohr, J.K., Sankey, O.F., Leinenweber, K., Somayazulu, M., and McMillan, P.F. (2000) Investigation of hardness in tetrahedrally bonded nonmolecular CO_2 solids by density-functional theory. *Phys. Rev. B*, **62**, 14685–14689.
71. Holm, B., Ahuja, R., Belonoshko, A., and Johansson, B. (2000) Theoretical investigation of high pressure phases of carbon dioxide. *Phys. Rev. Lett.*, **85**, 1258–1261.
72. Oganov, A.R., Ma, Y., Lyakhov, A.O., Valle, M., and Gatti, C. (2010) Evolutionary crystal structure prediction as a method for the discovery of minerals and materials. *Rev. Mineral. Geochem.*, **71**, 271–298.
73. Oganov, A.R., and Solozhenko, V.L. (2009) Boron: a hunt for superhard polymorphs. *J. Superhard Materials*, **31**, 285–291.

8
Pathways of Structural Transformations in Reconstructive Phase Transitions: Insights from Transition Path Sampling Molecular Dynamics

Stefano Leoni and Salah Eddine Boulfelfel

8.1
Introduction

In the practice of solid-state chemistry, structural phase transitions are fairly common events. Nonetheless a rationalization of the observed changes in symmetry pattern, and a deep understanding of the mechanisms are outstanding problems. Several approaches have been used to describe phase transformations based on kinetic, thermodynamic, and structural aspects [1]. The thermodynamic classification distinguishes between first- and second-order transitions according to the discontinuous behavior of quantities related to first or second derivatives of the free energy, respectively. Structural phase transitions with the second-order (continuous) character entail small atomic displacements and latent heat of a few calories per gram only. Additionally, the symmetries of the phases surrounding the transition are typically group–subgroup related. Reconstructive phase transitions on the contrary involve breaking of (large) parts of the bond scaffolding of the initial structure, and exhibit drastic changes at the transition, with large latent heat and hysteresis effects. Atomic displacements can be in the order of the lattice parameters or even larger, and no group–subgroup relation is found, between the symmetries of the phases. First-order phase transitions proceed by nucleation and subsequent growth of the new phase from the initial one. Different from continuous phase transitions, they imply coexistence of the transforming motifs. The discontinuity in some order parameter between the two phases is driven by lowering of the free energy as the new phase forms. However, close to the transition, the original phase remains metastable, and a fluctuation is needed for the formation of the new phase. Such a process responds to thermal changes, and depending on the height of the nucleation barrier, its rate may be slower or faster. In the former case, large deviations from equilibrium may be required to achieve transformation to the stable phase, which means that large hysteresis effects will be observed in the course of transformation. In this by nature complex field, the intellectually tantalizing task consists in elucidating mechanisms and in giving a face to intermediate configurations appearing along first-order phase transitions, and in solid–solid reconstructive processes in particular.

Modern Methods of Crystal Structure Prediction. Edited by Artem R. Oganov
Copyright © 2011 WILEY-VCH Verlag GmbH & Co. KGaA, Weinheim
ISBN: 978-3-527-40939-6

8.1.1
Shape of the Nuclei

The direct experimental observation of nuclei in first-order phase transitions is difficult, due to the small length and timescales characterizing such processes. In contrast with homogeneous nucleation in crystals, colloidal systems offer a better starting situation for observing nuclei [2], because of both, a larger size and a slower timescale. Their shape is determined to be nonspherical, on the average, and individual nuclei display rough rather than faceted surface. This represents a progress in the study of first-order phase transitions for at least two reasons. First, such investigations help clarifying the shape of the nuclei, giving access to some key parameters for understanding nucleation, like surface tension. Second, they underline the distinction between phase nucleation, which may span a long period of time and phase growth, which on the contrary can be very quick and trigger a rapid material transformation. The latter includes the possibility of a distinct structure in the nucleation and growth regions.

On the theoretical side [3–5], molecular dynamics simulations [6] offer in principle a powerful tool for investigating mechanistic issues, as all degrees of freedom can be explicitly considered. In practice, however, intrinsic difficulties are encountered.

In a recent work on the simulation of water crystallization into ice [7], the factor limiting the efficiency of molecular dynamics consists in long quiescence periods due to activation barrier much larger than $k_B T$ ($\Delta G^* \gg k_B T$). A lot of time is spent waiting for a particular configuration of hydrogen bonds to come into existence. The mutual arrangement of water molecules and the hydrogen-bonding scaffolding that are productive in terms of transformation into ice represent a very small subset compared to the very large number of network configurations that can be visited. On the timescale of MD simulations, the time window where transformation into ice is actually observed is very narrow, compared to the long waiting time (Figure 8.1). Therefore, the transformation is a rare event [6, 8].

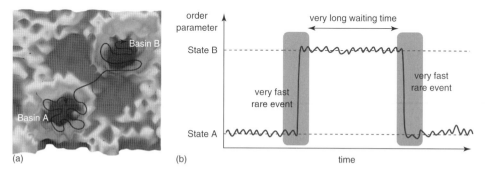

Figure 8.1 (a) Time evolution of a dynamical trajectory of a system connecting stable states A and B. (b) 2D projection: quiescence time spent in A, B is very long compared to the time needed to switch between states.

8.2
Transition Path Sampling Molecular Dynamics

The problem of rare events [9–13] in computer simulations arises from timescale disparity connected to the existence of high activation barriers (first-order phase transitions, chemical reactions, mass transport phenomena, and protein folding). If a simulation is started in basin A (Figure 8.1a), the system will spend most of its time in thermal agitation within it. Crossings to basin B, however, are rare and fleeting. Along this line, a natural disparity appears on the process timescale, slow and frequent vs. rare and quick. In straightforward molecular dynamics simulations, a basin can be escaped after a very long waiting time spent therein. This casual crossing is a far too fragile starting point for capturing details of the mechanism, and to filter out random noise effects, caused by orthogonal modes. A recent approach to tackle the rare event problem [14–17] is represented by the transition path sampling (TPS) method [9, 18–20].

The potential energy surface of a complex system is unlikely to be represented by saddle points alone. Instead, a multitude of points becomes relevant, some of which may still be stationary points, but many others not. Crossing a barrier between stable states may happen in a multitude of different ways, the transition path ensemble, beyond the idea of a single, well-defined transition path. The implementation of TPS consists in collecting such an ensemble. Therein, true dynamical trajectories are harvested in a Monte Carlo like sampling, which can be plugged on top of different, deterministic and stochastic simulation schemes [6]. The relevance of each path is weighed by its occurrence probability. The use of transition path sampling [21–26] is typically successful in systems characterized by two distinguishable stable (or metastable) long-lived states. A good case study for TPS is the pressure-induced phase transition B1–B2 in NaCl characterized by a double-well potential and a high activation barrier. On the contrary, the crystallization of water is expected to face many long-lived metastable conformations along the process, a situation less ideal for TPS.

8.2.1
First Trajectory

The machinery of TPS requires a first trajectory, where the reactive event is observed. There is no general recipe for modeling a first path in complex systems. For our purpose, the elucidation of transformation mechanisms in pressure-induced transformations in solids, we typically use a geometric/topological approach based on transforming periodic nodal surfaces [27–32] as illustrated in the following. This approach allows modeling a series of putative intermediate atomic configurations linking structure A to B. Instead of starting molecular dynamics runs from stable configurations, intermediates are propagated forward and backward in time until two trajectory segments give the transition (Figure 8.2).

In addition to a first trajectory, a means to delimit regions of initial and final states in the large space of configurations is needed. For this, an easy-to-use way

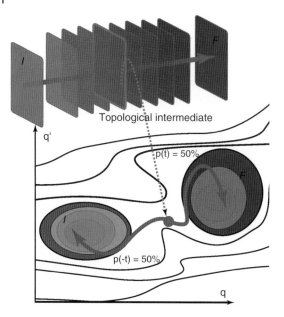

Figure 8.2 Construction of a first trajectory from geometric models. Upper part: the geometric model produces a dense set of configurations, intermediate between I and F, along a collective, one-dimensional order parameter (red arrow). The configuration corresponding to 50% probability of relaxing toward either I or F is selected as dynamical intermediate. From it, propagating in both directions of time, a first trajectory can be readily generated.

is the evaluation of a low-dimensional order parameter. This parameter can be very coarse within initial and final state basins, in order to smooth off equilibrium thermal fluctuations. The transition on the contrary must be reflected by a steep change of its value.

8.2.2
Trajectory Shooting and Shifting

TPS consists in gathering the transition path ensemble by the combination of two different dynamics performed in different spaces [9, 19, 20]. The core dynamics is performed in configuration space and produces trajectories by means of molecular dynamics runs. On top of it, a Monte Carlo walk in trajectory space is responsible of collecting the trajectories and distinguishing successful (A → B and B → A) from failed paths (A → A and B → B) [33, 34].

TPS implements two main operations on trajectories, shooting and shifting moves [35] (Figure 8.3). A new path $x^{(n)}(\tau)$ is generated from an existing (old) one $x^{(o)}(\tau)$ by randomly selecting a timeslice $x_t^{(o)}$ and applying momenta modifications. The perturbed configuration $x_t^{(n)}$ is then "shot off" in both directions of time. If the resulting trajectory still connects initial to final basin, it is accepted based on

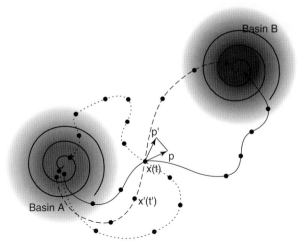

Figure 8.3 Shooting a new trajectory (dashed line) from an old one (solid line). Momenta modifications are introduced at a timeslice t(shooting move). Propagation in both directions of time may generate a successful new trajectory (dashed line) or a failed one (dotted line). If successful, a new shooting point x'(t') is chosen (shifting move) and the procedure is repeated.

the path probability [35]:

$$P_{acc}[x^{(o)}(\tau) \to x^{(n)}(\tau)] = h_A\big[x_0^{(n)}\big]h_B\big[x_\tau^{(n)}\big] \min\left[1, \frac{\rho(x_t^{(n)})}{\rho(x_t^{(o)})}\right]$$

with

$$h_{A,B}(x) = \begin{cases} 1, & \text{if } x \in A, B \\ 0, & \text{if } x \notin A, B \end{cases}$$

The distributions of position and momenta at timeslice t' for pathways $x^{(o)}$ and $x^{(n)}$ are $\rho(x_t^{(o)})$ and $\rho(x_t^{(n)})$, respectively. The acceptance/rejection criteria only depend on the shooting point, and can be easily translated into an algorithm [35]. The introduction of such a probability weighting scheme allows the system to cross-intermediate activation barriers in the order of $k_B T$ or lower.

The evolution from the initial trajectory is controlled by a sequence of shooting steps, iteratively enchained to each other, meant to collecting the transition path ensemble. For each subsequent step, the shooting point is different from the previous one (shifting), ensuring decoupling of the trajectories. To monitor the progress of trajectory generation, and to decide when a reliable mechanistic analysis is possible, a way of distinguishing between different stages of TPS is required. An incremental approach to trajectory shooting may be used to reach a regime, where the amount of momenta modifications represents a balanced rate between acceptance and rejection.

8.3
The Lesson of Sodium Chloride

Many of the alkali halides undergo a reconstructive phase transition from NaCl type (B1) to CsCl type (B2) under pressure. The prototypical compound, NaCl, is commonly used as a pressure standard, and has been the object of many investigations over the years [36–44]. NaCl crystallizes in the space group $Fm\bar{3}m$ at normal conditions (Figure 8.4a). It turns into a CsCl-type arrangement ($Pm\bar{3}m$) (Figure 8.4b) for pressure values above 30 GPa, with a large volume contraction and a large hysteresis [45].

In connection with orientation relations collected during diffraction experiments [46], attention has been payed to the problem of the crystallographic mapping of positions from one structure to the other. Outstanding are the models named after Bürger [47] and Hyde and O'Keeffe [48]. In the former, a contraction along the body diagonal and an expansion normal to it represent a possible connection between NaCl and CsCl. No changes in Wyckoff parameters are taking place and the transition is accomplished by strain alone. The latter consists of interplanar movements and shuffles of atoms in adjacent $(001)_{NaCl}$ in an antiparallel fashion. Therein, $[110]_{CsCl}$ is parallel to $[100]_{NaCl}$, while $[001]_{CsCl}$ is parallel to $[011]_{CsCl}$. This model involves changes in the Wyckoff structural parameters as well as in lattice modules. Both models are continuous variations from one structural motif to the other. A recent work enumerates transformation paths for the B1–B2 transformation in NaCl [36]. Therein, based on group-theoretical concepts, a list of as many as 12 paths is presented, and classified in terms of strain and activation energy. The basic idea is the identification of common subgroups between B1 and B2. The transformation path consists in varying the parameters between these two limiting points. The screening of transformation paths is then augmented by DFT-based total energy calculations.

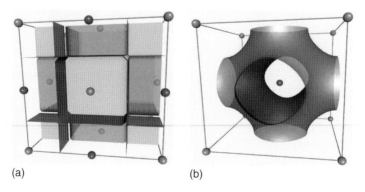

(a) (b)

Figure 8.4 (a) NaCl in the B1 structure type, $Fm\bar{3}m$ (b) NaCl in the B2 structure type, $Pm\bar{3}m$. The structural motifs are supplemented by a periodic nodal surface (PNS) to emphasize charge order and position separation.

Table 8.1 Possible geometric transformation paths from B1 to B2 in NaCl, adapted from [36].

No.	Space group symmetry	Energy (eV)
1	$R\bar{3}m$ (166)	0.077
2	$C2$ (1) (5)	0.91
3	$C2$ (2) (5)	0.90
4	$Cmc2_1$ (36)	2.08
5	$Pmmn$ (59)	0.10
6	$P2/c$ (1) (13)	1.17
7	P/c (2) (13)	1.04
8	$P2_1/m$ (11)	2.15
9	Pc (7)	0.92
10	$C2/c$ (15)	0.58
11	$P\bar{1}$ (2)	0.40
12	$Iba2$ (45)	1.33

After relaxation of intermediate configurations, possible further lowering of the symmetry is investigated, reflected into a reduced activation barrier. Based on the DFT scoring, the Bürger mechanism (Table 8.1, No. 1) appears as the best candidate for the B1–B2 phase transition in NaCl, followed by the Hyde and O'Keeffe model (Table 8.1, No. 5) [36].

8.3.1
Simulation Strategy

Each mechanism (Table 8.1) codes distinct, quasi-orthogonal atomic movements. Distinguishable intermediate configurations (for a choice of the parameters halfway along the transformation path) may appear in connection with a particular deformation. However, as long as the reaction coordinate remains undisclosed, the question of intermediate configurations visited during phase transition must remain open. Nonetheless, each model, although potentially biased by the modeling approach, represents a way of connecting B1 and B2.

This feature can advantageously be used in the derivation of a first trajectory as required by the TPS simulation scheme. Without overdriving the system, finding a dynamical pathway between B1 and B2 is far from straightforward. However, using geometric models as starting points, the search for a dynamic trajectory corresponding to a transition at the critical pressure and temperature becomes tractable [49–51].

8.3.2
Topological Models

Instead of using group–subgroup mappings of Wyckoff positions, we model transformation paths by transforming triply periodic nodal surfaces (TPNS), which have shown to provide a deeper insight into the organization of crystalline matter, and to allow for an unequivocal classification of net types [27–32]. The reciprocal space approach implements short Fourier summations to define a family of surfaces, according to the formula [27]:

$$f(x,y,z) = \sum_{h,k,l} \|S_{hkl}\| \cos(2\pi(hx+ky+lz) - \alpha_{hkl})$$

where $h = (hkl)$ and $x = (xyz)$ are vectors in reciprocal and real space, respectively. S_{hkl} is a geometric structure factor and α_{hkl} is the corresponding phase. The surface corresponding to $f(x, y, z) = 0$ is called PNS [27]. Triply periodic functions are oriented and partition space into interpenetrating labyrinths (commonly two). In the case of NaCl, the Na^+ and Cl^- ions can be placed on different sides of a PNS calculated from the (eight) cubic permutations of $h = (111)$ with no phase restrictions ($\alpha_{hkl} = 0$). The resulting surface, called F* [28] has the symmetry $Fm\bar{3}m$ (Figure 8.4a). Similarly, Na^+ and Cl^- ions in the high-pressure cubic modification can be placed in different labyrinths of a PNS calculated from the (six) cubic permutations of vector $h = (100)(\alpha_{hkl} = 0)$, which generates a surface of space group $Pm\bar{3}m$ (Figure 8.4b), called P* [28]. On choosing different isovalues, the surface becomes a collection of disconnected "bubbles" centered on the Na^+ and Cl^- sites, on the positive ($f(x, y, z) > 0$) and on the negative side ($f(x, y, z) < 0$) of the surface, respectively (Figure 8.5).

The cubic surfaces P* and F* are the starting points for the definition of a geometric model connecting the NaCl to the CsCl-type structure [30]. The choice of a common cell with a constant number of atoms and the periodicity of the model ensure commensurability of the two phases. After transformation of the reflections of the new setting of the cell, the transition can be formulated as a migration from one structure to the other along a coordinate s providing weighted linear mixing of the two functions:

$$f_{AB}(x,y,z) = sw_A f_A(x,y,z) + (1-s)w_B f_B(x,y,z), s \in [0,1]$$

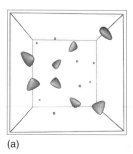
(a)
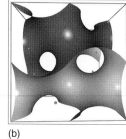
(b)
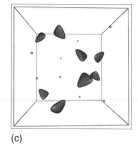
(c)

Figure 8.5 Different values of $f(x, y, z)$: (a) negative values, disconnected surface. (b) $f(x, y, z) = 0$, PNS, continuous surface. (c) Positive values, disconnected surface. The surface is calculated from the permutations of (110, $\frac{\pi}{2}$) in $I4_132$ [27].

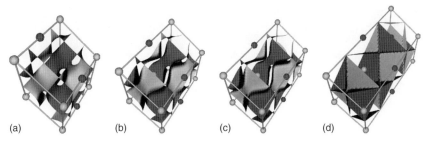

Figure 8.6 The Burger mechanism in terms of periodic surfaces. (a) P* and (d) F* surface representing the B2 and B1 structures, respectively, (b) and (c) are intermediate configurations.

As the PNS separates positive and negative "charges" from each other, for particular isovalues of the starting function, $f_A(x, y, z)$, the surface is a collection of bubbles enclosing positive and negative charges, respectively. The weighting factors w_A and w_B are chosen such that the bubbles remain disconnected for each value of s. Variation of the mixing factor, s, results in a concerted atomic movement, periodic at each stage, and continuous in the atomic displacements. The degree of mixing of the limiting functions may be interpreted in terms of a reaction coordinate. For each value of $s \in [0, 1]$, a different configuration of the atoms is connected. The cell setting can be chosen rhombohedral (hexagonal setting), applying the transformation matrix $(101, \bar{1}11, 0\bar{1}1)$ to the cubic NaCl cell. The set (111), eight vectors, splits into (101), six vectors, and (003), two vectors. The (101) set alone generates the topology of the P* surface. Both surfaces collapse at sites (0, 0, 0) and (1/2, 0, 0), and the only free parameter is the ratio c/a, which can be used as reaction coordinate. This reproduces the Bürger model (Figure 8.6).

Transformation of the cubic reflex sets into a common orthorhombic cell is the starting point for the description of the Hyde and O'Keeffe model in terms of periodic surfaces. The origin of the F cell has to be shifted by $(\frac{1}{4}, \frac{1}{4}, 0)$, whereby the phases of the vectors (111) and (11$\bar{1}$) change from 0 to π. The origin of the P cell is shifted by $(0, \frac{1}{2}, 0)$, which only affects the phase of reflex (010, $\alpha_{010} = \pi$). The surface describing the NaCl phase collapses around sites (3/4, 1/4, 1/4) and (1/4, 1/4, 1/4) for positive or negative choices of the isovalue, respectively, while the sites are (1/4, 1/4, 0) and (3/4, 1/4, 1/2) for the CsCl-type structure. The linear mixing results in a continuous and synchronous movement of the Na^+ and Cl^- sites as a function of the degree of mixing, s. Atoms in adjacent $(100)_{NaCl}$ layers displace along [110] in an antiparallel fashion. Hereby each atom undergoes a displacement of 1/8 of the face diagonal of the NaCl unit cell (Figure 8.7).

The modeling approach defines a one-dimensional collective coordinate, s. For each path the transition states correspond to a single value s_{TS}. We refer to the transition state as the configuration along the collective coordinate s, which equals probability of relaxing toward either NaCl or CsCl. If $p(NaCl)$ represents the probability of forming NaCl, the expression $p_{NaCl} = 0.5 = p_{CsCl}$ defines the transition state configuration. To derive the transition state, configurations obtained

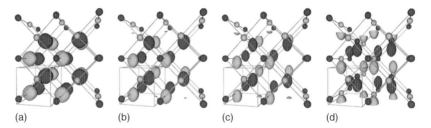

Figure 8.7 Hyde and O'Keeffe mechanism modeled with periodic surfaces. (a) NaCl (B1) type structure. The surface $f(x, y, z)$ is a collection of disconnected bubbles centered on Na (red) and Cl (yellow). (d) CsCl (B2) type structure. (b) and (c) are intermediate produced by sliding (110) layers.

from geometrical models in the range $s \in [0, 1]$ where propagated in molecular dynamics simulations at 300 K [30]. For this, random velocities were assigned to the atoms. For different sets of initial velocity distributions, an averaged value of s can be found. For the mechanism in Figure 8.7, the intermediate structure and the probability profiles are displayed in Figure 8.8.

8.3.3
Combining Modeling and Molecular Dynamics Simulations

The derivation of models based on geometric/topological modeling represents an efficient method for the generation of first trajectories. Far from trivial, this first step is crucial for successful transition path sampling molecular dynamics

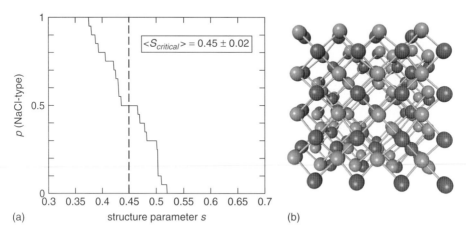

Figure 8.8 (a) Probability of finding the NaCl-type structure after relaxation of a starting configuration derived from geometrical models as a function of the collective parameter s. (b) Atomic configuration corresponding to $s = 0.45$.

simulations [30, 52]. The system is let free to evolve toward one attractor, B1 or B2, in unprejudiced molecular dynamics simulation runs [52]. On reversing the sign of the time coordinate, the other attractor can be reached with a finite probability. This provides a simple, yet effective way of generating a first trajectory at the experimental values of temperature and pressure. Furthermore, many initial trajectory types can be generated, from different transformation models.

8.3.4
The Mechanism of the B1–B2 Phase Transition

The typical evolution of a TPSMD run shows a quick departure from the features inherited from the model, that is collective atomic movements and concerted mechanisms toward a regime where the reconstruction is initiated locally, followed by growth of the stable phase. A representative snapshot sequence is given in Figure 8.9, where a regime characterized by nucleation and growth clearly sets in.

The nucleus is represented by a few or even a single atom (Figure 8.9a, blue spot), followed by phase growth, which is carried by antiparallel layer shuffling in the orthogonal direction to the initial layer displacements.

On the average, the mechanism corresponds to the Hyde and O'Keeffe mechanism (Figure 8.10), with the difference that the global antiparallel layer movements of the model are resolved into the displacement of single layers under formation of an interface between region of B1 and B2 structural motifs (Figure 8.11). This represents an important difference with respect to previous models. Within the

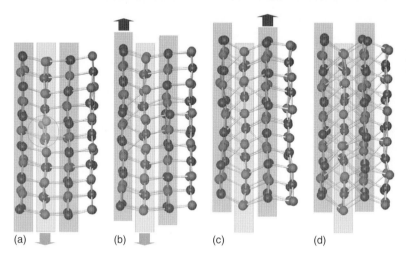

Figure 8.9 (a) (B1) Initial displacement of one atom out of its equilibrium position. (b) Layer shifting following the initial displacement. (c) Propagation of the transformation front by antiparallel layer shuffling (red and yellow arrows). (d) Complete reconstruction of the lattice into B2.

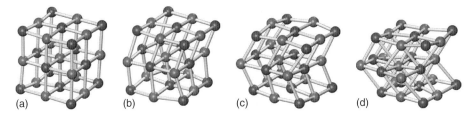

(a) (b) (c) (d)

Figure 8.10 Global, average mechanism as found from MD simulations. The mechanism corresponds to the Hyde and O'Keeffe model. It implies (110) layers shuffling. A portion of the simulation box corresponding to the standard unit cell choice for B1 is shown. (a) B1; (d) B2; (b) and (c) intermediate stages.

common subgroup approach, the possibility of such an intermediate is by design lifted, as the collective movements do not allow for any interface to set in.

The interfacial region extends over three layers (Figure 8.11a), perpendicularly to the direction of growth, and is "infinite" in the other two. Considering the structural pattern inside the interface, particularly the polyhedron of the seven-coordinated atom, the correspondence to α-TlI is striking [53]. The possible intermediate role of α-TlI (B33) in the B1–B2 phase transitions has already been postulated [54, 55], however always as a proper intermediate, that is a structural intermediate involving the whole structure, like B1 → B33 → B2. On the contrary, the B33 motif appears as an interface, in our case [52].

There is thus no proper intermediate of the B33 type (in fact none was detected by experiments). Instead, B1 and B2 coexist because of the interface. Different from static calculations [56, 57], the scenario disclosed by TPSMD simulations is the one of a stable phase nucleating and growing from the metastable one, under formation and propagation of an interface [58].

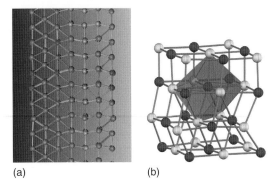

(a) (b)

Figure 8.11 (a) Interface between regions of B2 and B1 structural motifs. (b) α-TlI structural motif present in the interfacial region. Notice the characteristic coordination polyhedron.

The above-mentioned advantage of generating many different first trajectories from the modeling approach is useful in testing the stability of the mechanism on changing the initial conditions. Starting, for example, the simulation runs from a Bürger mechanism (Figure 8.6), a final Hyde and O'Keeffe mechanism (Figure 8.7) is obtained. Apart from further supporting the mechanistic analysis, this resolves the issue raised by the evaluation of the potential energy barrier [36] in connection with symmetry models. The Bürger mechanism is ruled out, although it appeared as the favored one from static calculations. The second best, the Hyde and O'Keeffe mechanism, is the winning mechanism instead. This shows the necessity of moving the mechanistic investigation away from static calculations.

8.3.5
Crossing the Line: NaBr

The pressure-induced B1–B2 phase transition is well documented for K, Rb, Cs halides, while it is absent for Li halides [59]. Na halides occupy an intermediate positions: NaF transforms from B1 into B2 around 27 GPa, NaCl around 29 GPa. NaBr and NaI, on the contrary, have been found to undergo a transformation to an undetermined, noncubic structure [60, 61], which only recently was determined to be of the α-TlI-type (B33) [62]. While this underlines the role of the B33 structural motif on the transformation path connecting B1 and B2, it offers at the same time an interesting challenge to the simulation strategy presented above. In fact, a simulation run could be started from a trajectory connecting B1–B33, with the aim of investigating its mechanism. However, this would not be very enlightening concerning the real intermediate role of the B33 structure type.

In the spirit of the modeling approach which is used to start the TPSMD runs, a trajectory connecting B1 and B2 for NaBr was used instead, and the simulation was performed at the experimental pressure, that is 30 GPa. The challenge for the simulation is represented by the fact that neither the mechanism nor one of the transforming structures, the B2 structure in this case, are supposed to represent the real system. For NaCl, monitoring the change of the coordination number ($CN = 6(B1) \leftrightarrow CN = 8(B2)$) represents a good order parameter for monitoring the sampling progress. In this simulation a slightly changed definition was used:

$$CN_{B1,B2}(x) : \begin{cases} \leq 6 \Rightarrow x \in B1 \\ \geq 7 \Rightarrow x \in B2 \end{cases}$$

This choice of the order parameter allows the growth of the B33 structure type. In the course of trajectory sampling, trajectories initially still connect B1 and B2 according to the mechanism of Hyde and O'Keeffe with B33 motifs appearing at the interface. On further path sampling, these B33 motifs grow over many layers (Figure 8.12c). In the final regime, B2 has been completely replaced by B33, which thus represents the configuration NaBr transforms to under pressure.

This simulation strategy proves that the topological models, in connection with TPSMD, can be used in a predictive way. For difficult problems, where a phase is only tentatively determined, such a strategy may provide important insights.

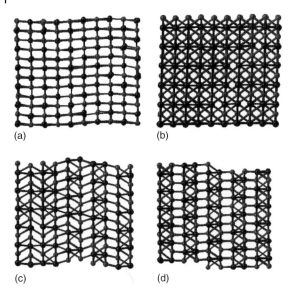

Figure 8.12 B1–B33 transformation in NaBr. (a) Initial B1 configuration. (b) B2 configuration of the initial trajectory regime, B1–B2. (c) Intermediate configuration along a trajectory connecting B1 and B2 with large regions of B33 structure. (d) Final trajectory regime, with B33 replacing the former B2.

The calculations performed on NaBr show that during TPSMD iterations both the intermediate regime (the mechanism) and one of the structures surrounding the phase transition can be optimized (Figure 8.12c). This means that the simulation setup can be used also as a predictive tool, for exploring the free-energy landscape in search of new polymorphs (Figure 8.13). For this, an initial trajectory can be spanned between two known modifications, and the trajectory propagated in TPSMD. Given the iterative way of the optimization process, the method can be less attractive with respect to time efficiency, compared for examples to metadynamics [63]. Nonetheless, since it is the whole transition path that is being optimized, more details about the true preferences of the system can be collected, in terms of mechanism and preferred intermediate configurations. This approach may, for example, disclose the crossing over to a different mechanistic regime, and the reasons for the disfavor of the former.

8.4
The Formation of Domains

In continuous transitions, in the presence of group–subgroup symmetry relationships, the formation of domains can be related to the symmetry-lowering process along two principal lines [64, 65]:

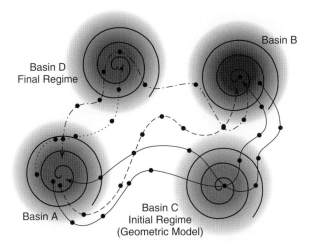

Figure 8.13 Evolution of trajectory regimes for a simulation involving change of one of the trajectory basins. This scheme applies to the results obtained for NaBr. A, B, and D correspond to B1, B2 and B33, respectively.

- If the symmetry-lowering step involves changes of the point-group symmetry (*translationsgleich*, t), twins or multiple twins can be expected.
- If translations are lost (*klassengleich*, k), antiphase domains are built.

In this context, symmetry, and representation theory in particular, constitute powerful tools for a deep understanding of domain structures. The number of domain orientations (domain types) is determined by the group to subgroup index.

If a unique way of representing symmetry changes is missing, in the general case of noncontinuous phase transitions, the number and the type of domains remain however undetermined. The fact that paths can be constructed, which go through common subgroups, does not impose any constraint on number or type of domains that can be expected, or predict any domain formation at all.

To shed further light on nucleation, growth, and domain formation, we turn to KF. Similarly to NaCl, it transforms from B1 to B2 under pressure [66, 67]. Different from NaCl, it shows an intrinsically higher nucleation density, which allows for observing many nucleation centers during TPSMD, and permits the study of domain formation in a multicenters, multidomains situation [68].

After trajectory sampling, the overall mechanism is found to be of the same type as found for NaCl, namely the Hyde and O'Keeffe mechanism. However, for a simulation volume comparable to NaCl, many nucleation centers can be observed, which grow into domains. In the overall B1–B2 transformation mechanism, layers are shifted by half of a K–K distance. However, the layers are not moved as an entity, but by subsequently sliding columns of anions or cations, following rules that we shall describe now.

The coordination number (CN) of the F^- ions is tracked by a color code. In a B2 crystal viewed along [001] (Figure 8.14 and, CN = 8, transparent blue), a column

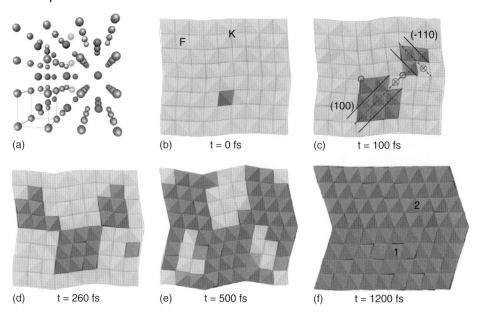

Figure 8.14 B2–B1 transformation in KF. (a) Initial columnar displacement of fluoride ions (bright green) in the B2 structure (K grey, F green). (b) Formation of a first nucleation center. (c) Growth of the first nucleation center, and formation of a second center. (d) and (e) Further growth into the final B1 structure. (f) Two domains are formed, separated by a mirror.

of fluoride ions is displaced and promotes the formation of a small island of B1 structure (CN = 6, green). After about 100 fs another nucleation center appears, (Figure 8.14c). The two nuclei grow by subsequent shifting of adjacent ion columns along [001].

In Figure 8.14c, the directions of the up/down columnar displacements of the fluoride ions are indicated by dotted and crossed circles, respectively. K^+ ions are not shown explicitly in this representation; however, the planes of adjacent K^+ and F^- columns, which are shifted in the same direction, are indicated as lines. Note the different orientation of the sliding planes in both phase domains. The growth of the two B1 domains hence leads to a frustrated contact region, indicated by blue circles. Further growth of both domains results in an interfacial layer (Figure 8.14f). After completion of the phase transition this interface represents a (100) mirror plane separating two B1 regions.

For the shifting of columns different directions are allowed by symmetry. However, temporarily coexisting nuclei were always found to involve parallel or antiparallel moves; mutual orthogonal orientations were not observed. A B1 nucleus originating from an initial [001] columnar displacement may involve the shuffling of (110) and ($\bar{1}$10) planes with equal probability. Different nuclei may hence result in phase domains derived from different initial shiftings. The compatibility of the initial moves determines the formation of twin boundaries

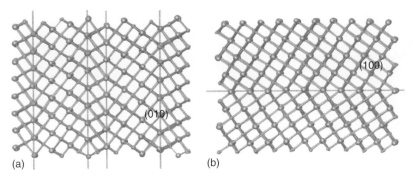

Figure 8.15 Different final domain structures resulting from different pattern of nucleation and growth. Different mirror planes separate the domains, (a) (010), (b) (100).

upon fusion of nucleation fronts. For many nuclei, many different domains with different extension can thus be formed.

On fusing nucleation fronts, smooth interfaces result, with distinct atomic configurations. Therein, the atoms (K in this case) are in a sixfold coordination. However, not in an octahedron, but in a trigonal prism instead. The mirror planes are formed instead of driving or constraining the transition. Thanks to the good sampling of configurations (ergodicity [6]), different mirror planes are formed in turn, i.e., (010) or (100) (Figure 8.15).

8.5
Structure of the B2–B1 Interfaces

The columnar shifting of ions in KF accounts for the different nucleus morphology with respect to NaCl or KCl, where layers are displayed instead. In NaCl, the interface between B1 and B2 during phase growth shows motifs of the B33 structure (α-TlI). In KF we find a quick phase growth in the direction of the initial, columnar ion displacement, similarly to what was observed in NaCl. In KF, however, the transition from the nucleation event into a situation of domain growth is less abrupt, with no unique sharp interfaces (like for NaCl). Many interfaces are formed, instead. Therein, B16/B33 motifs can be identified (Figure 8.16, B16 is the orthorhombic GeS-type structure).

To invoke the existence of B33 motifs is a useful step for better approaching the description of the difficult intermediate situation of coexistence of structural motifs (B1 and B2) during B2 structure reconstruction. However, there is no evidence for any role of group–subgroup relationships, like they may be constructed along the structural sequence B2–B33–B1. The symmetries of the boundary structures, B1 and B2, both allow for an exchange of the labeling of the atoms, without change of the structures. In this respect, the transition state is very different from the initial and final configurations. Only one chemical species is selected to move first, F^-. B2 nucleates on sliding columns of F^- ions in B1, followed by K^+. The exchange

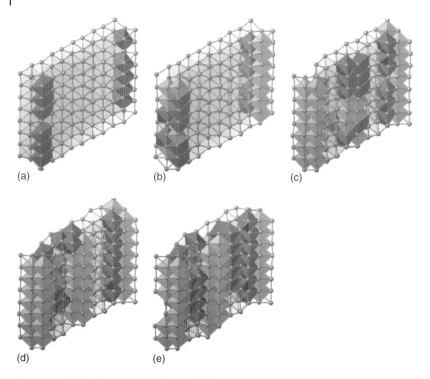

Figure 8.16 Interfaces between B1 and B2 formed during crystal reconstruction. (a) Initial columnar shiftings (an orange polyhedron means partial shifting of the central fluoride ion). (b) Islands of B1 (yellow), and further growth, (c)–(d). (e) Larger regions of B1 structure separated by interfaces. The fraction of B2 is very reduced.

of K^+ ions by F^- ions implies an energy cost of 0.04 eV per ion [68]. The black and white symmetry of the B1 and B2 phases is completely lifted in the intermediate region, due to nucleation and growth.

The different behavior of K^+ and F^- ions may be explained by the chemical concept of hardness and softness [68]. The chemically hard F^- ions exhibit a very small electronic polarizability. As a consequence, the expected way to accommodate F^- ions in response to local lattice fluctuations is to move.

On the other hand, K^+ ions can do for a much larger electronic polarizability; it is softer. Hence apart from moving, the potassium ions have hence an additional means of responding to local stress. This softness is expected to account for the observed difference in K^+ and F^- ionic motion during phase nucleation processes.

To better capture the role of ionic hardness and softness on local reconstruction patterns, the evolution of the final morphology of potassium halides was investigated on varying the halide moiety, from F^- over Cl^- to Br^- [69]. With respect to the ionic hardness/softness ratio, potassium represents the softer species in KF, whereby in KBr the roles are inverted, K^+ being relatively hard with respect to Br^-.

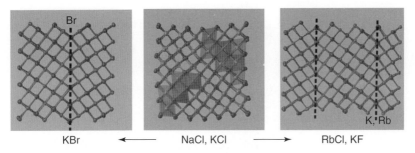

Figure 8.17 Different final morphologies as a consequence of a different ionic hardness/softness ratio. In KF domains are separated by mirror planes (dashed line) and K^+ ions occupy the sites at the interface (in RbCl, Rb sits at the interface). In KBr, Br^- ions, the softer species, are placed at the interface, instead.

KCl occupies an intermediate position. As a consequence, domains are formed for the combinations with the largest difference in ionic hardness/softness, KF and KBr. KCl does not show any pronounced tendency to form domains within a given volume, like it was the case for NaCl. KF and KBr break down into domains instead. While their morphology appears similar at first sight, as in both cases domains are separated by mirror planes, there is an important difference that shows up upon closer inspection. The chemical species that is occupying the interface – the energetically most problematic place in the structure – is always the softer one.

In KF, K^+ is sitting at the interface, on the geometric place of the mirror plane separating the domains (Figure 8.17). In KBr the analogous site is occupied by the softer Br^- ions. Clearly, apart from geometry and mechanics, there is a tremendous role of chemical reactivity. Nucleation is initiated by fluoride ions movements in KF, by K^+ ions shift in KBr. In situation of coexisting structural motifs, the interfaces take on the task of propagating growth, and are the reactive places in the transforming material. Especially at these places, there is a marked difference in the way cations and anions are rearranged, and influence the final interface morphology.

8.5.1
Domain Formation in RbCl

In RbCl distinguishable nucleation centers appear asynchronously on the timescale of the simulation. The overall, average mechanism is still the Hyde and O'Keeffe mechanism, similarly to what was observed in NaCl. Adjacent {110} layers are displaced in an antiparallel way, for the reconstruction to complete. However, looking at Figure 8.18, the form taken by the transformation in RbCl is very remote from the collective, concerted way of understanding a classical mechanism. The Hyde and O'Keeffe mechanism picks one layer set out of the {110} manifold. However, in case of distant, noncorrelated nuclei, slidings of different layers can

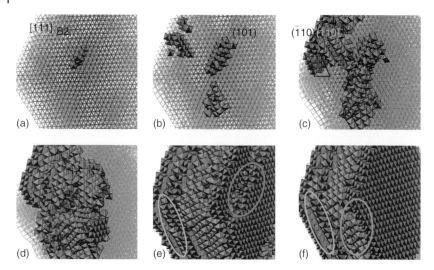

Figure 8.18 B2–B1 reconstruction in RbCl. Regions of B1 structure are marked with (pink) coordination polyhedra around Rb. (a) and (b) Initial nucleation events are spatially apart and uncorrelated. (b) and (c) Growth of larger regions of B1. Local formation of twin domains separated by trigonal prismatic polyhedra (green region), (c). (c)–(f) Further growth and fusion of B1 regions, under formation of smooth interfaces (blue ellipses), rough interfaces (green ellipses) and no interfaces (red ellipses). (Please find a color version of this figure on the color plates.)

be activated, (101) besides (110) for example. Different regions of the material may thus express a different version of the Hyde and O'Keeffe materials, with respect to the sliding layers.

In Figure 8.18, an RbCl crystal transforms from B2 into B1 (marked by pink polyhedra). From the very beginning, different orientations are shown by distant nuclei (Figures 8.18a–c). Their uncorrelated growth is apparent by inspecting the regions of B2 structure between the islands, which remain undistorted at this stage. In Figure 8.18c, next to regions of (110) layer sliding, twin domains are built. Like in KF, a mirror plane is forming between regions of (110) and $(\bar{1}10)$ layer displacements. Nuclei in this region are highly correlated, and the regions of B2 structure between them rapidly consumed. Furthermore, the interface formed between them is smooth, and coincide with the geometric place of a mirror plane. In the polyhedral representation, this interface is characterized by trigonal prismatic polyhedra (Figure 8.18c, green region). This does not need to be the case for distant nuclei. In Figures 8.18c–e, growth fronts are coming into contact and fusing (Figures 8.18e and f), under formation of different interface morphologies. Twin domains are separated by smooth interfaces (blue ellipses in Figures 8.18e and f), and are associated with correlated nucleation events. Rough interfaces (Figure 8.18f, green circle) result on the contrary from the fusion of propagation fronts associated with different sliding layers, (110) and (101) for example, which are not related by a mirror symmetry operation. The interface structure may locally

8.5 Structure of the B2–B1 Interfaces

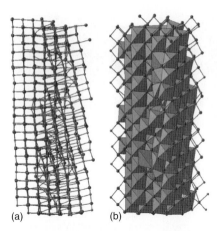

Figure 8.19 Details of a rough interface. (a) Skew arrangement of lattices in the rough interfacial region. (b) Polyhedral representation, rotated with respect to (a). Notice the distorted polyhedra and the voids.

(a)   (b)

further relax upon contact, leading to a complicated final geometry. A closeup of a portion of the interface in this region is displayed in Figure 8.19.

A third way of fusion of growth fronts implies the formation of no interface (Figures 8.18e and f, red ellipse), when the fronts are associated with layer shuffling with identical plane index, (110) and (110), for example. The fronts consume the B2 region between them and fuse voidlessly.

8.5.2
Liquid Interfaces in CaF_2

The many examples provided so far have demonstrated the central role of interfaces during crystal reconstruction. The fronts of propagation have been shown to bear peculiar structural pattern, distinct from the transforming crystal and from the ordered pattern the latter is reconstructing into. Often, familiar atomic arrangements can be recognized therein, which can be related to known structure types, like α-TlI. Depending on the extension and dimensionality of the interface regions, the comparison with a known structural motif can be more or less legitimate.

Transformation processes between liquid and solid phases are phenomena of central importance in nature. The most fundamental transition of this kind is represented by crystallization from the melt. While this is one of the main synthetic pathways in solid-state synthesis with a liquid–solid interface, a more peculiar phenomenon is reflected by superionic conductors, a fascinating blend of solid and liquid states that arises when only parts of a compound become liquid.

In CaF_2, the fluoride sublattice is known to melt on raising temperature. The liquid–solid interface is within the crystal, during this premelting step, between the still solid sublattice (Ca^{2+}) and the liquid one (F^-). The remarkable technological potential of ion-conducting materials based on distinct transport effects has been reviewed [70]. Distinctly enhanced ionic conduction can be achieved at interface region of $CaF_2 - BaF_2 - CaF_2$ sandwich structures [70]. The key of understanding this anomaly is again in the different structural places represented by the interfaces. Therein, bulk properties are lifted, and "strange" phenomena, like enhanced mass

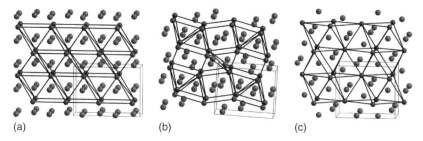

Figure 8.20 (a) CaF$_2$ in the fluorite-type structure and (c) in the cotunnite-type structure. (b) Geometric intermediate along the symmetry branch $Fm\bar{3}m \Rightarrow Immm \Rightarrow Pnma$.

mobility, or incomplete charge balance, can be accommodated [71]. The use of pressure to cause lattice reconstruction in superionic compounds is thus expected to reserve surprises, particularly in connection with interfaces. In the following, the coexistence of fusion and condensation steps during pressure-induced lattice reconstruction in CaF$_2$ is summarized [72].

The polymorphism of CaF$_2$ encompasses two fundamental structural types, the low-pressure face-centered cubic fluorite structure of space group $Fm\bar{3}m$, for which CaF$_2$ is the prototype compound, and the high-pressure PbCl$_2$ type for the orthorhombic polymorph cotunnite (Figure 8.20) of space group $Pnma$ [73, 74]. In the fluorite structure, Ca^{2+} ions are in a cubic close-packed ccp arrangement, the fluoride ions occupy all the tetrahedral interstitial sites. In the cotunnite structure, the array of Ca^{2+} ions is hexagonal close-packed hcp [75], half of the fluoride ions exhibit tetrahedral coordination, the other half is placed off center in the ideally octahedral voids of hcp, with fivefold coordination. Ca^{2+} are eightfold coordinated in the fluorite structure, while the coordination number increases to nine in the denser cotunnite structure. The fluorite–cotunnite phase transition occurs in the range 9.5–20 GPa [76] with the cotunnite phase retransforming to the fluorite phase on releasing pressure. The high-pressure polymorph can be quenched from higher temperature, 300° C, nonetheless the crystallinity of the obtained material is poor.

To start the calculations in the TPSMD scheme, a first trajectory is needed. To derive it, we take advantage of the possibility of relating the space groups of the fluorite-type structure and of the cotunnite-type structure by different group–subgroup symmetry paths [77]. We chose the right branch (Figure 8.21) through $Immm$. The derived mechanism contains the symmetric intermediate configuration of Figure 8.20b, featuring trigonal prisms of Ca^{2+} ions, which are formed by the fusion of adjacent tetrahedra on increasing the puckering of the layers of the cubic structure.

The way of reconstructing the Ca^{2+} close-packed hexagonal pattern coded in the starting route (Figure 8.20) disappears during the early iteration steps, and a different mechanism sets in. The signature of the onset of the cubic-to-orthorhombic phase transition is an enhanced mobility of the fluoride ions. While the arrangement of the Ca^{2+} ions remains as in the fluorite structure, the F$^-$ ions can switch

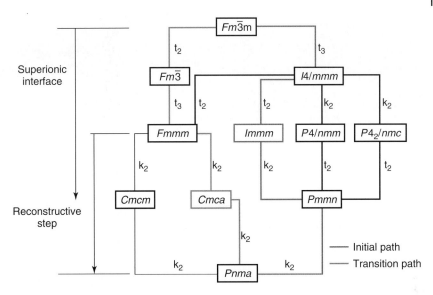

Figure 8.21 Symmetry paths connecting the space groups of fluorite type (top) and cotunnite type (bottom). Different intermediate space groups and configurations are possible, divided into two main branches, a tetragonal (right) and an orthorhombic one (left).

between different tetrahedral voids, with excursions as long as 650 pm. Anion mobility is achieved through the creation of Frenkel defects, by episodic occupation of octahedral voids (Figures 8.22a and b). In this less dense part of the structure the reconstruction of the Ca sublattice is initiated (Figure 8.22c). Displacements of portions of (100) Ca layers along [011] with respect to the fluorite structure remove the central octahedral void. This is accompanied by the formation of an octahedral void with a different orientation. A displacement within an adjacent (100) Ca^{2+} layer rearranges another octahedron (Figure 8.22d). The fluoride ions inside these octahedra move off-center in pyramidal coordination.

The mechanistic analysis indicates a difference between F^- and Ca^{2+} ions with respect to ion mobility. The paths left behind by fluoride ions (Figure 8.23) have been traced by illuminated lines over a period of 1.2 ps [78]. F^- ions that are still in the fluorite configuration display short traces, while in the interfacial region they exhibit a markedly enhanced mobility reflected into a longer trace. Furthermore, chains are formed that propagate diagonally in the box. The ionic jumps are frequent in the regions where octahedra are reconstructing (red polyhedra) and reach out to neighboring regions, where the reconstruction is about to start. While the overall F^- ions displacements follow crystallographic directions, local jumps are uncorrelated, such that this region is liquid like. The setup of an interfacial region corresponds to locally melting the fluoride sublattice. Upon propagation of the phase front (green polyhedra), the liquid-like interface is shifted such that an enhanced mobility characterizes this region only. This picture

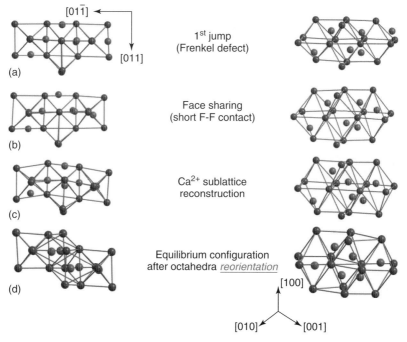

Figure 8.22 Frenkel defects and lattice reconstruction (portion of the simulation box, two different views). (a) and (b) Occupation of octahedral voids in the fluorite structure type. (b) and (c) Beginning of lattice reconstruction around Frenkel defects (red), and formation of distinct octahedral voids. (d) Final arrangement in the cotunnite-type structure, with reoriented octahedral voids.

is intrinsically different from a conventional solid–solid transformation. In the latter, the displacement paths of rearranging atoms are well defined in both direction and length. Here, F$^-$ ions abandon their initial configuration to occupy a final one, percolating through an interface. For quantitative calculations on ionic conduction, the existence of continuous paths between region of bulk or bulk-like properties and interfaces is a critical knowledge [79]. The simulation are very precise in giving a face to the interface, and especially in showing how defects are accommodated within or just at the boundaries of the interfacial region, for the ions to rearrange.

8.6
Domain Fragmentation in CdSe Under Pressure

The investigations on the kinetics of first-order structural transitions have turned to nanocrystals as prototypes for single-domain phase transformation [80–83]. Many insights have been collected regarding critical lengths [81] and possible transformation mechanisms. On monitoring shape changes, the activation volume

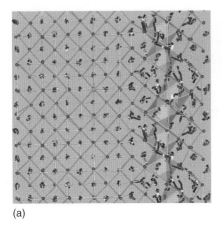 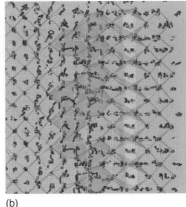

Figure 8.23 (a) In the reconstructing, interfacial region octahedral voids are rearranged and occupied by fluoride ions (transparent red polyhedra). Only in this region enhanced anion mobility is observed. (b) Shift of the phase propagation front. Anion mobility is enhanced only in the interface region (green polyhedra) between cubic (left) and orthorhombic atomic pattern. The displacements of fluoride ions are traced in a time window of 1.2 ps.

can be directly investigated and mechanisms can be argued [82, 83]. However, a precise atomistic understanding of mechanisms and domain formation, and a reliable extrapolation from nano to bulk materials are still an open issue. In the following, we are considering the evolution of domain morphologies and shape in bulk CdSe under the effect of pressure, both from theory and by performing experiments. The mechanistic issues of CdSe transformation have been an open problem for many decades. In nanocrystal experiments, particularly on CdSe and ZnO, structural transformations are traced back to a single nucleation event and to simplest kinetic [80]. CdSe is a wide-gap semiconductor that has found extensive use in optical application for its rich set of effects in the nanoregime. Under normal conditions it crystallizes in the wurtzite structure type (B4). Applying moderate pressure (2.5–3.5 GPa), it transforms into rocksalt structure type (B1) [84]. As a third polymorph, B3 (zincblende structure type) does exist, albeit metastable. Evidence of possible coexistence of structural motifs of B3 and B4 has been collected from experiments on nanocrystals [82, 83], bulk CdSe [85], and from mechanically manipulating samples of B4 structure [86]. Irregularities in resistivity measurements may reflect mixed composition [87]. However, no insight into the atomistic origin of such coexistence has been proposed. A combination of B4 and B3 regions in a single material represents a means of achieving the most simple domain boundaries, without locally disrupting the lattice or introducing vacancies. Nonetheless, first principles calculations conclude at the energetic closeness of the B4 and B3 with B4 representing the ground state structure [88]. To elucidate the formation of domains under pressure and to back-trace domain fragmentation to nucleation events, we have performed TPSMD simulations [9, 89] on the B4–B1 reconstructive phase transition. While any initial trajectory connecting B4 to B1

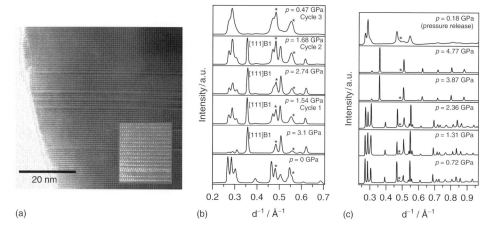

Figure 8.24 (a) HR-TEM before probe pressurization. Large domains (about 20 nm) separated by lamellar insets are visible. The wurtzite pattern is enlarged in the framed inset. (b) X-ray powder diffraction collected *in situ* during three pressurization cycles. In regions of phase coexistence between $p = 1.5$ and $p = 3.1$ GPa signal coarsening is indicative of domain fragmentation. (c) X-ray powder diffraction from *in situ* synchrotron measurements during a single pressurization run up to 4.77 GPa followed by pressure release. Notice the loss of signal structuring. Reflection lines marked with stars are due to the metal gasket of the diamond cell.

can in principle be considered, we have used a geometric model connecting B3 and B1, bypassing B4 on purpose. This strategy corresponds to starting the simulation in a regime delimited by metastable modifications, B1 and B3, and converging the simulation runs toward a stable regime.

8.6.1
B4–B1–B4 Transformation

In Figure 8.24, an HR-TEM image of the starting material is displayed. Therein extensive domains of wurtzite structure type are visible with tiny lamellar insets. Complete transformation of the starting material into B1 is achieved above 3.5 GPa, with a hysteresis offset of about 1 GPa with respect to equilibrium pressure [85, 89]. Upon release of pressure, the appearance of the diffraction pattern is remarkably different from the starting one. A broadening of the peaks is evident, as well as a strong perturbation of the relative line intensities, especially in the manifold around 0.3 Å^{-1} (Figures 8.24b and c). This pattern is characteristic of the final product, and a slower decompression or a different pressurization profile does not appreciably alter it, which suggests a lower limit for domain fragmentation. The initial B4 material is thus very different from the final product. This hints at a structurally diverse scenario of (sub)nanodomains that originate in the high-pressure regime.

TPSMD allows to shed light on the region of phase coexistence (Figure 8.25). In the B1 structure matrix (Figure 8.25a) initial nuclei of B4 structural pattern

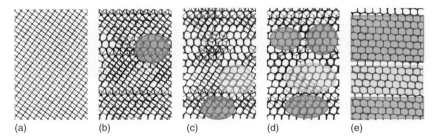

Figure 8.25 (a) Initial B1 configuration. (b)–(d) Nucleation of B4 (purple) and B3 (yellow) motifs, respectively. (e) Final lamellar arrangement. (Please find a color version of this figure on the color plates.)

start forming (Figure 8.25b, purple circles), followed by regions of B3 structural motifs (Figure 8.25c, yellow circle), which grow between already defined B4 regions, in a more confined region of space. Notice the dominance of B4 regions at this point. The final (sub)nanodomains result from further growth from this initial configuration (Figure 8.25e). No domain recombination is observed during growth. Instead both B3 and B4 are originating from the B1 structure and do not imply intra-layer rearrangements. Intermediate patchwork configurations of B1, B3, B4 (Figures 8.25c and d) characterize the later transformation stages, and slow down the nonetheless complete transformation to the final void-free four-connected structural motif (Figure 8.25e). Lamellae of B3 phase are clearly visible between more extended B4 regions. This lamellar structure, where wurtzite is the dominating component ($\frac{1}{4}$ B3 to $\frac{3}{4}$ B4, on average) is typical for the lower pressure range inside the hysteresis, here $p = 2.5$ GPa. In this region the simulations are able to provide detailed atomistic insights on phase coexistence, which are of otherwise difficult determination.

The local nucleation events leading to B3 and B4 structural motifs are subsuming similar but distinctive mechanistic pattern. Three-layers portion of regions about to growing to B4 and B3 domains, respectively, are shown in Figure 8.26. The final layer sequence aAbBaA – distinctive of the hexagonal stacking of B4 – is formed by displacing Cd^{2+} ions between Se^{2-} layers (Figure 8.26, B4 mechanism), whereby only three of the initially 6 Se^{2-} ions remain in the next-neighbor (n–n) coordination sphere of cations. For B3 (Figure 8.26) of cubic layer sequence aAbBcC, cations are shifted in a similar manner, but do conserve 4 of the initially six anions in their nearest neighbor coordination sphere (B3 mechanism).

Both displacements are taking place within (001) layers perpendicular to $[0001]_{B4}$ of the final wurtzite regions. They contain a diffusive and a displacive component, the former being more pronounced in the B4 mechanism, while the latter is more pertinent to the B3 mechanism. B3 formation appears more martensitic-like, possibly reflecting the confined growth condition. Their alternation suggests a strategy to locally minimize strain, which would be very large on full layer sliding.

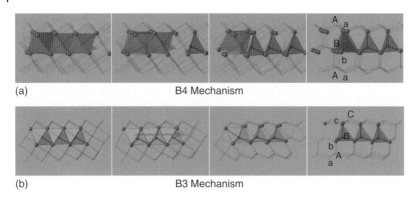

Figure 8.26 (a) Diffusive-like mechanism characterizing the growth of B4 regions (B4 Mechanism). (b) Displacive-like character connected to the growth of B3 regions (B3 Mechanism).

Accordingly, different region may respond to changed external pressure in a different way.

To better understand the role of local transformation pattern and morphology change details nominal pressures can be varied. In going from lower ($p = 3.5$ GPa) to slightly higher pressure (Figure 8.27c, $p = 3.8$ GPa) the ratio of B3 to B4 lamellae has increased, from B3/B4 $\approx$ 0.8 to B3/B4 $\approx$ 1.0, with B3 and B4 layer thickness becoming similar (Figure 8.27b). In the HR-TEM image of Figure 8.27a, a lamellar structure is apparent, with alternating B3 lamellae of different thickness involving 3–5 layers. Here we present two images from different probes exposed to different loadings. The sample boundaries are typically dominated by B3 structure motifs. In Figure 8.27d a lamellar arrangement with a dominating internal region of B3 structure is shown. Both simulations and HR-TEM sample investigations do agree on the features, that is (a) lamellar appearance of the sample and stacking direction, and (b) size and distribution variation of lamellae, with pronounced role of B3 on increasing pressure. Owing to the difference in the mechanisms connected to nucleating B3 or B4 patterns, their response to pressure enhancement is also different. B3 regions do on average grow faster, not particularly because of an enhanced single-domain growth kinetic. This rather reflects a change in the nucleation pattern as an increment in the number of events leading to B3 domains. While at lower pressure ($p = 2.5$ GPa) already $\frac{1}{2}$ of the final B4 regions has formed at times of B3 nucleation onset, at higher pressure ($p = 3.5$ GPa) only $\frac{1}{3}$ of B4 is built when B3 starts growing. B4 also grows from many tiny lamellae propagating from as many nucleation centers. On average B3 growth is thus enhanced under the effect of pressure. Analogously to nanocrystals individual domains in the bulk (B3 or B4) are connected to distinct generating events. Different from nanocrystals, a nucleation pattern rather than a single nucleation event promotes phase transformation (Figure 8.27).

8.6 Domain Fragmentation in CdSe Under Pressure

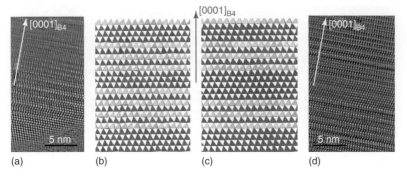

Figure 8.27 (a) HR-TEM images of lamellar domains on pressurized samples. (b) and (c) Final configuration from MD simulations in the higher pressure regime between $p = 3.5$ GPa and $p = 3.8$ GPa. (d) HR-TEM images of another sample with larger domains of the B3 structural motif.

8.6.2
Defects

Lamellar domains of B3 structure are separated by B4 lamellae. Additionally, within a B3 domain skew three-layer insets of B4 structural motifs can be formed. As for the B3 defects, this does not require introducing vacancies in the lattice, as it is apparent from inspecting Figure 8.28. In the polyhedral representation (Figure 8.28b) local distortions are visible. This type of defect can be recognized in the HR-TEM image of Figure 8.28a. An inset of B4 structural motif is visible, that separates region within a B3 domain and terminates inside a B4 domain. Within the B4 region the defect does not cause any dislocation in the atomic pattern (Figures 8.28a–c), as the incidence angle with the (001) planes of about 70° is preserved. Only very locally, at the beginning of the defect, changes are visible as distorted six-membered rings. In a recent work [85], tilting angles of about 5° between final B1 ⟨100⟩ planes and B4 [0001] have been reported. Additionally, deviations within (001) appear. Our simulations and experiments support a correspondence between B1 [111] and B4 [0001], consistent with experiments on compressing nanocrystals. Nonetheless, the skew defects introduce a reorientation of B4 planes perpendicular to [0001], which form an angle of about 5° with B1 [001] (Figures 8.28c and d). It is fascinating to observe how different direction (mis)alignments can be accommodated in the same vacancy-free lattice of fourfold coordinated atoms, without invoking a scenario of a severely mechanically perturbed sample.

8.6.3
The Lesson of CdSe

The joint investigation on CdSe, by theory and experiments [90], shows very clearly that geometric/crystallographic arguments alone cannot afford the degree of complication represented by real phase transitions. The fragmentation into

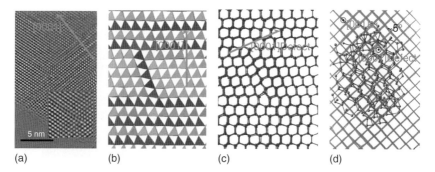

Figure 8.28 (a) HR-TEM image of defects, magnified in the inset. (b) and (c) Defects of B4 structural motifs from MD simulations. The [0001] direction of the B4 defect is indicated. (d) Alignment of [0001] of the B4 defect with respect to the initial [001] of the B1 structure. An offset of about 5° is visible.

domains under pressure demonstrates how specific the details of phase transformations can be. Without a clear picture of nucleation pattern, neither domain formation, nor defects can be fully understood. Apart from this, the possibility of performing simulations at nominal pressure values close to experiments offers a different way of understanding the interplay of theory and experiment. Simulations are performed hand-in-hand with experiments, filling in missing parts, and helping a consistent picture of the chemistry/physics of the system to form. As for today, there is no general theoretical framework for transformation mechanisms. For the time being, we understand the judicious application of appropriate numerical strategies to understanding critical phenomena as a way to access chemical details of these difficult and exciting processes.

8.7
Intermediate Structures During Phase Transitions

The elucidation of the existence of intermediate structures in phase transitions is a central issue in modern solid-state sciences. One example is the regular publication of novel high-pressure polymorphs in simple systems [91–93]. Silicon for instance shows a large number of polymorphs, some of which have been discovered in recent years, like *Cmca* Si between *sh* and *hcp* forms [94]. Sometimes, it is the appearance of additional diffraction lines in high-pressure experiments, which is indicative of the existence of a distinct structural motif. In other situations, anomalies in the phonon spectrum close to critical thermodynamic values may indicate intermediate stages, like in GaN or ZnO. The interest in intermediates is on the line of the search for novel polymorph of a material, which may offer improved properties. With respect to band gap tuning, the high-pressure modification of GaN would perform better than the stable wurtzite structure. Unlike in AlN, the high-pressure modification of

GaN is metastable at ambient conditions and reverts to the stable form on releasing pressure. So the very basic question on what determines metastability for a given composition (what Alivisatos calls [95] : the rules of metastability) is posed here.

In many geometric models, intermediates may be enforced by the form given to the order parameter and the symmetry coded therein suggests a direction. However, the search for intermediates corresponds to a free-energy scan and needs more articulated approaches. Metadynamics is, for example, a better solution [63]. This method was recently demonstrated to correctly capture intermediates in the silicon systems, in the correct sequence as a function of pressure [96]. Another approach is a mechanistic investigation in search of transformation regimes which may host an intermediate.

8.7.1
Intermediates Along the Pressure-Induced Transformation of GaN

Group-III nitrides (GaN, AlN, and InN) are semiconducting materials with extensive applications [97, 98]. Among III–V, II–VI, and IV–VI semiconductors, GaN stands out for very short bond length, high ionicity, small equilibrium volume, low compressibility, high thermal conductivity, high melting temperature, and large energy band gap. As a consequence, the fabrication of crystals or films is rather difficult. Cation substitution allows control over the wavelength of emission because of the differences in band gaps (GaN 3.4 eV, AlN 6.3 eV, and InN 1.9 eV) [99]. GaN (AlN and InN) crystallizes in the wurtzite structure (B4 type). Under pressure the B4 structure transforms into the rocksalt structure (B1 type) [100, 101] with a large hysteresis effect (37–54 GPa). The very fast and destructive character of the transition reflects a first-order thermodynamics that complicates the identification of the real transition pathway from experiments. Consequently, obtaining a reliable scenario for the transformation by determination of the detailed atomistic mechanism has been a challenging problem for several years, for both experiments and theory. Experimentally, the first attempt to describe the WZ–RS transition as a continuous deformation was proposed by Croll [102], based on his experiments of ultrasonic pulse-echo measurements of elastic properties. Later, using X-ray diffraction and optical absorption under pressure, Tolbert and Alivisatos [103] proposed a series of models picturing the transition as deformation of chair/boat type patterns into 2 × 3 rectangles. Another transformation path going through a face-centered orthorhombic intermediate configuration was suggested from shock experiments [104].

The initial stage of the WZ–RS transition has been investigated using picosecond time-resolved electronic spectroscopy leading to a new transformation model where a face-centered tetragonal structure was identified as a necessary step for the transition [105]. Limpijumnong and Lambrecht [106] proposed an orthorhombic path for this transition, crossing the h-MgO type structure (Figure 8.29, blue path). This is realized by changing the c/a ratio from $c/a = 1.63$ to $c/a = 1.2$ and the relative spacing u of the Ga and N (or z coordinate of one ion) sublattices from

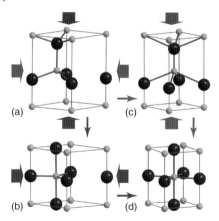

Figure 8.29 Two possible paths for the transformation from the (a) wurtzite (B4, $c/a = 1.63$) to the (d) rocksalt (B1, $c/a = \sqrt{2}$). Shear strain deformation via (b) h-MgO intermediate ($c/a = 1.63$, blue path). Deformation via a (c) tetragonal intermediate ($c/a = 1.74$, red path).

$u = 0.377$ to $u = 0.5$. Subsequent perpendicular compression changes the ratio c/a to $\sqrt{2}$ and leads to the B1-type structure.

Saitta and Decremps [107] presented a different distortion referred to as tetragonal path (Figure 8.29, red path). In a common monoclinic cell, the γ angle reduces from 120° to 90°, while ions move horizontally toward the center of a square pyramid formed by five counter-ions. On lowering the c/a ratio, the ions reach the center of the pyramid base (B1 structure). The prediction of such tetragonal deformation is based on the softening of shear (c44 and c66) and phonon E_2^{low} modes. The possibility to combine the atomic movements discussed above allows for another set of viable paths [108, 109].

The crucial point in this system, and the simulation challenge, is the distinction between the suggested paths, in terms of overall mechanisms and intermediate configurations crossed along the transition. For a convincing simulation approach, it is thus crucial to prepare initial trajectories that contain the one or the other intermediate, and also include initial trajectories which avoid such intermediates (Figure 8.30), as blind probes so to speak. In the course of simulations, the disfavor or the advantage of a trajectory regime should result from the Monte Carlo walks of TPSMD. In case of a disfavored way of transforming B4 into B1, the reasons for the disappearance of some features can be investigated and understood.

TPSMD simulations performed on GaN are very explicit concerning the nature of the intermediate. Trajectories containing the hexagonal intermediate (Figure 8.29b) are not probable and the corresponding regime not visited at all.

In the initial regime, sixrings in wurtzite structure are initially compressed along [120] such that additional Ga–N contacts result. These tentative nucleation centers regress again and the system is rearranged into a symmetric configuration related to the hexagonal wurtzite structure but slightly compressed along [120]. Such a

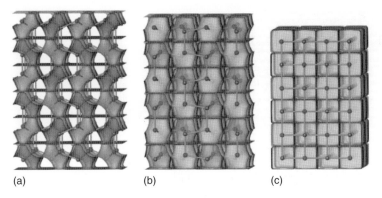

Figure 8.30 Model transformation mapping the network of (a) B4 onto the network of (c) B1. Ga (gray) and N (red) atom sets are separated from each other by a PNS (gray/green manifold). The latter develops perpendicularly with respect to the shortest Ga–N connections. Snapshot (b) represents an intermediate between (a) and (c).

configuration corresponds to the symmetric intermediate structures prepared via geometric modeling. In the course of TPS iterations configurations corresponding to a tendency to stay longer in a coordination number equal to five clearly emerges. After full trajectory decorrelation, this tendency becomes a pronounced feature ruling the hexagonal intermediate out.

The most favorable transition regime proceeds via reconstruction in (001) plane giving birth to nucleation centers along [001] (Figure 8.31, 0 ps). These centers grow and gain more volume (Figure 8.31, 3–4 ps) converting the initial hexagonal wurtzite structure into an intermediate tetragonal structure with fivefold coordinated ions (Figure 8.31, 5 ps). This intermediate is similar to the high-pressure modification of tin phosphide, SnP [110] and GeAs [111]. The second step in the transition mechanism implies a compression along [001] moving ions to lock in their final positions in the rocksalt structure (Figure 8.31, 5–8 ps). This is accompanied with an increase of coordination number of ions from five to six.

The time resolution of picosecond time-resolved electronic spectroscopy is around 100 ps, and only in particular cases helpful in identifying a possible intermediate, like for CdS in shock wave experiments [112]. One of the main experimental difficulties is the strong overlap of the tetragonal structure peaks with those of wurtzite and rocksalt during X-ray powder diffraction (XRD) data refinement [113]. The time resolution of our simulations on the contrary allows for a clear identification of distinct events. The nucleation and growth sequence, wurtzite-tetragonal then tetragonal-rocksalt, clearly reflects a two-step transformation mechanism where the tetragonal intermediate does not only simply link the limiting structures in a crystallographic way, as described by Saitta and Decremps, but stands as a metastable one, reflecting a different energetic profile of the transition path.

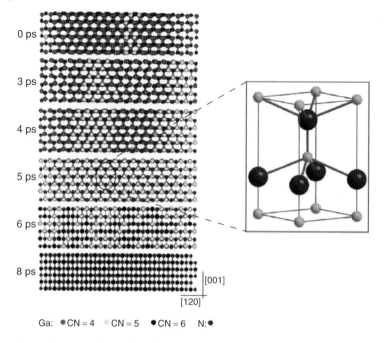

Ga: ● CN = 4   ◌ CN = 5   ● CN = 6   N: ●

Figure 8.31 Snapshots taken from a representative trajectory of the B4–B1 transformation of GaN. The trajectory features an intermediate structure (5 ps) with fivefold coordinated ions. This intermediate is tetragonal (inset) and ions occupy the core of square pyramids.

8.7.2
Polymorphism and Transformations of ZnO: Tetragonal or Hexagonal Intermediate?

The polymorphism of ZnO is similar to the one of GaN. It exhibits a B4–B1 transformation under pressure, and has a zincblende B3 structure as a metastable phase [114–117]. The transition is reversible and shows a pronounced hysteresis (2–10 GPa) [118–120]. Fraction of the B1 phase can be quenched [121–123]. Additionally, imposing external loadings, other polymorphs have been pointed out [124]. Similarly, the intermediates mentioned for GaN represent possible routes for the B4–B1 transition in ZnO. This offers a nice opportunity of addressing the problem, whether a mechanism connecting B4 and B1 can be considered fully transferable over chemically different systems (from GaN to ZnO), and which is the role of chemistry.

Several experiments tried to explore the polymorphism of ZnO [125–129], as well as the proposed models for the B4–B1 transformation [130–135]. The two intermediate configurations in question are the hexagonal iH and the tetragonal iT (Figure 8.32).

The investigation of phonon anomalies is a valuable tool in this system. The ZnO wurtzite phase shows Γ point optic modes of $A_1 + 2B_2 + E_1 + 2E_2$ symmetry

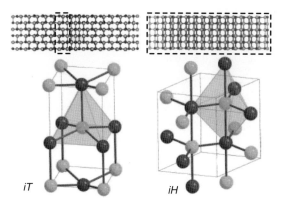

Figure 8.32 Intermediate tetragonal iT and hexagonal iH motifs appearing during TPS simulation runs. The former is characteristic of the B4/B1 interface along the displacing layers. The latter results from a more collective displacement mode along $[001]_{B4}$.

character (A_1, E_1, and E_2 are Raman active while A_1 and E_1 are infrared active) [133]. The transition model B4-iH-B1 suggests the optic A_1 and E_2^{high} modes to be affected by the first step B4-iH. However, high-pressure Raman spectroscopy investigations show no instability of these modes during the transition [133]. The only Grüneisen parameter that shows a negative value corresponds to E_2^{low} mode [126]. In fact, this instability, together with the softening of shear modes under pressure may involve the tetragonal path B4-iT-B1 described for GaN (Figure 8.29). However, up to date, neither intermediate (iT or iH) has been characterized in experiments.

Either hexagonal or tetragonal intermediate are formed by a different synchronization of the variation of relative spacing u between cations and anions sublattices with $(001)_{B4}$ layers shearing.

If the compression along $[001]_{B4}$ occurs during (or after) the $(001)_{B4}$ layers reconstruction, the hexagonal intermediate is ruled out and the system crosses over directly to the tetragonal intermediate. Otherwise a compression along [001] prior to the shearing leads to the hexagonal intermediate. With respect to the layer shearing modes which are productive toward lattice reconstruction into B1, the details of the transition are similar to GaN. However, the simulations disclose a scenario of fluctuation of u close to phase transition. Indeed, Liu et al. showed [122] that the increase in the structural internal parameter u is observed up to ~9 GPa with a maximal value $u = 0.43$ at around 5.6 GPa, well below the critical pressure 10 GPa and before the appearance of the B1 phase. As the rocksalt structure starts forming, the parameter u shows a sharp decrease to its initial value of 0.38.

This reversible pretransitional effect reported in experiments is nicely reproduced in our simulations, as reflected in the volume profiles on Figure 8.33.

Different from GaN, there is no proper intermediate that is visited along B1–B4 phase transition. The tetragonal motif of the GaN intermediate is present in ZnO,

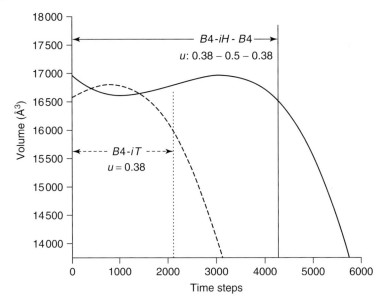

Figure 8.33 Volume profiles of transition paths visiting the *iH* (plain line) or bypassing it (dashed line). A different evolution of the internal parameter *u* distinguishes the two paths, as well as a larger volume increase for the B4-*iH*-B1 path. The latter is disfavored by a pressure increase and does not represent a necessary step for the transition.

however only at the B4/B1 interface. Additionally, besides the tetragonal motif, a hexagonal intermediate does appear. The latter represents a pretransitional step, and is not productive for the transition.

The B4–B1 transition in ZnO is accompanied by a large volume collapse (∼20 %). Considering the profile of the volume evolution associated with paths containing the B4-*iH* pretransitional step (Figure 8.33, plain line) or directly crossing over to B1 via the *iT* intermediate (Figure 8.33, dashed line), the former allows for a slightly higher maximum. The associated [001] compression implies a comparatively smaller orthogonal expansion. The cell and volume evolution anisotropy, difficult to trace in experiments, can nonetheless be investigated in the simulations. The occurrence of the *iH* structural pattern results in longer transition paths (25 ps instead 13 ps). Increasing pressure above 10 GPa shortens the trajectories and alters completely the pretransitional step B4-*iH* in favor of a B4-*iT*-B1 path. This supports an interpretation of the *iH* intermediate as controlled by a fluctuation of the *u* parameter close to phase transition and corresponding to different responses of the system to external pressure loadings. The weighed probability of each mechanism deduced from our simulations is in agreement with the recent high-pressure experiments on ZnO investigating the influence of deviatoric stress in connection with both possible tetragonal and hexagonal intermediates [136].

8.8
Conclusions

The nucleation and growth scenarios emerging from performing TPS molecular dynamics simulations clearly indicate the need of a definitive departure from modeling reconstruction mechanisms based on symmetry/geometry approaches only. Our simulations disclose fundamentally different atomistic landscapes of reconstructive phase transitions. Even if an average mechanism can in principle be defined, such a move may tarnish the important distinction between nucleation and growth. As it was shown at many places, the times and modes of material nucleation may be dramatically different from those of material growth. Already with NaCl, the initial nucleation triggers layer displacements and contributes setting up a thin interface, which propagates orthogonally to the initial nucleation directions. In KF and RbCl, the final material morphology is rooted in the initial nucleation pattern, and reflects the set of nucleation directions in the initial, yet untransformed material.

In a different class of materials, superionic conductors like CaF_2 the distance from a geometry-designed mechanism has further deepened. Therein not only interfaces are playing a central role, but one sublattice has even become liquid. Constitution of the interfaces and degree of disorder are key parameters for transformation critical lengths. In CaF_2 thin interfaces are associated with the liquid state of fluoride ions. This is fundamentally different from a way of understanding a mechanism with respect to one particular mode of moving atoms between initial and final configurations. There is a lot more in between, in terms of liquid-like states, interfaces, disorder, which in turn causes variation of Debye lengths, and so on. All this remains inaccessible if the modeling or simulation approaches are not able to account for the proper transformation length scales and times, and for the articulated configurations in-between.

The question, whether a particular configuration can appear along the path connecting initial and final configurations can only be answered by a theory, which is able to asses critical lengths, and to collect the true preferences of the system, at values of pressure and temperature corresponding to experiments. What is postulated from first principles approaches, based on unit cell calculations, may not be able to survive in the true transformation landscape, or may shrink down to the size of an interface.

In connection with the study of domain fragmentation in CdSe and domain formation in RbCl, the idea of a more or less extended topotaxy was pointed out. Therein the occurrence of chemical reactions, and the different reactivity of interfaces were observed. Metastable interfaces brings about distinct diffusion behavior of the chemical species involved. As a consequence, stable structures emerge as a result of the occurrence of the transformation mechanism and of chemical reactions within the phase-growth fronts. The focus is thus shifted on local reactions that are taking place at interstitial sites and interfaces, which are distinct places in the solid material. The role of symmetry is recast into more appropriate boundaries, and the need to abandon group–subgroup relationships for reconstructive phase transformations is emphasized. Instead, the role of chemistry in driving patterns of nucleation, domain formation, and the evolution of metastable interfaces and intermediates is clearly emerging. Therein the phase-formation sequence and its interplay with chemical reactivity controlling the final material morphologies become firmly rooted from assumption-free model studies. All in all, a perspective is emerging of bringing together computer simulations and solid–solid phase transitions, in a way that the chemical intuition is supported by the atomistic resolution of the computer models.

References

1. Chandra Sekhar, N. V. and Govinda Rajan, K. (2001) *Bull. Mater. Sci.*, **24**, 1.
2. Gasser, U., Weeks, E.R., Schofield, A., Pusey, P., and Weitz, D. (2001) *Science*, **292**, 258.
3. ten Wolde, P.R., Ruiz-Montero, M.J., and Frenkel, D.J. (1996) *J. Chem. Phys.*, **104**, 9932.
4. ten Wolde, P.R. and Frenkel, D.J. (1998) *J. Chem. Phys.*, **109**, 9901.
5. ten Wolde, P.R., Ruiz-Montero, M.J., Frenkel, D.J. (1999) *J. Chem. Phys.*, **110**, 1591.
6. Frenkel, D., Smit, B. (1996) *Understanding Molecular Simulation: From Algorithms to Applications*, Academic Press, San Diego.
7. Matsumoto, M., Saito, S., Ohmine, I. (2002) *Nature*, **416**, 409.
8. Allen, M.P. and Tildesley, D.J. (1987) *Computer Simulation of Liquids*, Clarendon Press, Oxford.
9. Bolhuis, P.G., Dellago, C. and Chandler, D. (1998) *Faraday Discuss.*, **110**, 421.
10. Zhou, H.-X. and Zwanzig, R. (1991) *J. Chem. Phys.*, **94**, 6147.
11. Anderson, J.B. (1973) *J. Chem. Phys.*, **58**, 4684.
12. Bennett, C.H. (1977) *Algorithms for Chemical Computations*, (eds R.E. Christoffersen), 63, Washington, DC, Amer. Chem. Soc..
13. Chandler, D. (1978) *J. Chem. Phys.*, **68**, 2959.
14. W. Ren, W.E. and Vanden-Eijnden, E. (2002) *Phys. Rev. B*, **66**, 052301.
15. W. Ren, W.E. and Vanden-Eijnden, E. (2005) *J. Phys. Chem. B*, **109**, 6688.
16. Elber, R., Ghosh, A., Cardenas, A., and Stern, H. (2004) *Adv. Chem. Phys.*, **126**, 93.
17. Passerone, D., Ceccarelli, M., and Parrinello, M. (2003) *J. Chem. Phys.*, **118**, 2025.
18. Pratt, L.R. (1986) *J. Chem. Phys.*, **85**, 5045.
19. Dellago, C., Bolhuis, P.G., Csajka, F.S., and Chandler, D. (1998) *J. Chem. Phys.*, **108**, 1964.
20. Dellago, C., Bolhuis, P.G., and Chandler, D. (1998) *J. Chem. Phys.*, **108**, 9236.
21. Bolhuis, P.G., Chandler, D., Dellago, C., and Geissler, P.L. (2002) *Ann. Rev. Phys. Chem.*, **53**, 291.
22. Dellago, C., Bolhuis, P.G., and Geissler, P.L. (2002) *Adv. Chem. Phys.*, **123**, 1.
23. Dellago, C., Chandler, D. in (2002) *Molecular Simulation for the Next Decade*, (eds P. Nielaba, M. Mareschal, G. Ciccotti), 321 Springer, Berlin.
24. Dellago, C., *Handbook of Materials Modeling*, (ed S. Yip), (2005) 1585 Springer, Berlin.
25. Dellago, C. in *Free energy calculations: Theory and Applications in Chemistry and Biology*, (eds A. Pohorille, C. Chipot), (2006) 265 Springer, Berlin.
26. Crooks, G.E. and Chandler, D. (2001) *Phys. Rev. E*, **64**, 026109.
27. von Schnering, H.G. and Nesper, R. (1991) *Z. Phys. B*, 407.
28. von Schnering, H.G. and Nesper, R. (1987) *Angew. Chem. Int. Ed. Engl.*, **26**, 1059.
29. Leoni, S. and Nesper, R. (2000) *Acta Crystallogr. A*, **56**, 383.

30. Leoni, S. and Zahn, D. (2004) *Z. Kristallogr.*, **219**, 345.
31. von Schnering, H.G., Zürn, A., Chang, J.-H., Baitinger, M., and Grin, Y. (2007) *Z. Anorg. Allg. Chem.*,, **633**, 1147.
32. Grin, Yu., Wedig, U., Wagner, F., von Schnering, H.G., and Savin, A. (1997) *J. Alloys Comp.*,, **255**, 203–208.
33. Landau, D.P. and Binder, K. (2000) *A Guide to Monte Carlo Simulations in Statistical Physics*, Cambridge University Press, Cambridge.
34. Metropolis, N., Rosenbluth, A.W., Rosenbluth, M.N., Teller, A.H., and Teller, E. (1953) *J. Chem. Phys.*, **21**, 1087.
35. Dellago, C., Bolhuis, P.G., and Geissler, P.L. (2006) *Lect. Notes Phys.*, **703**, 349.
36. Stokes, H.T. and Hatch, D.M. (2002) *Phys. Rev. B*, **65**, 144114.
37. Recio, J.M., Pendas, A.M., Francisco, E., Flores, M., and Luana, V. (1993) *Phys. Rev. B*, **48**, 5891.
38. Sowa, H. (2000) *Acta Crystallogr. A*, **56**, 288.
39. Nunes, G.S., Allen, P.B., and Martins, J.L. (1998) *Phys. Rev. B*, **57**, 5098.
40. Nga, A. and Ong, C.K. (1992) *Phys. Rev. B*, **46**, 10547.
41. Watanabe, M., Tokonami, M., and Morimoto, N. (1977) *Acta Crystallogr. A*, **33**, 284.
42. Stokes, H.T., Hatch, D.M., Dong, J., and Lewis, J.P. (2004) *Phys. Rev. B*, **69**, 174111.
43. Sims, C.E., Barrera, G.D., Allan, N.L., and Mackrodt, W.C. (1998) *Phys. Rev. B*, **57**, 11164.
44. Martín Pendás, A., Luaña, V., Recio, J.M., Flórez, M., Francisco, E., Blanco, M.A., and Kantorovich, L.N. (1994) *Phys. Rev. B*, **49**, 3066.
45. Li, X. and Jeanloz, R. (1987) *Phys. Rev. B*, **36**, 474.
46. Watanabe, M., Tokonami, M., and Morimoto, N. (1977) *Acta Cryst. A*, **33**, 294 ; Blaschko, O., Ernst, G., Quittner, G., Pépy, G., and Roth, M. (1979) *Phys. Rev. B*, **20**, 1157.
47. Bürger, M.J. in *Phase Transformations in Solids*, (eds R. Smoluchowski, J.E. Mayers, W.A. Weyl), (1951) 183 Wiley, New York.
48. Hyde, B.G., O'Keeffe, M. in *Phase Transitions*, (eds L.E. Cross), (1973) 345 Pergamon, Oxford.
49. Smith, W. and Forester, T.J. (1996) *Mol. Graphics*, **14**, 136.
50. Fumi, F.G. and Tosi, M.P. (1964) *J. Phys. Chem. Solids*, **25**, 31.
51. Melchionna, S., Ciccotti, G., and Holian, B.L. (1993) *Mol. Phys.*, **78**, 533.
52. Zahn, D. and Leoni, S. (2004) *Phys. Rev. Lett.*, **92**, 250201.
53. Helmholtz, L. (1936) *Z. Kristallogr.*, **95**, 129.
54. Tolédano, P., Knorr, K., Ehm, L., and Depmeier, W. (2003) *Phys. Rev. B*, **67**, 144106.
55. Toledano, P., Knorr, K., Ehm, L., and Depmeier, W. (2003) *Phys. Rev. B*, **67**, 144106.
56. Catti, M. (2003) *Phys. Rev. B*, **68**, 100101.
57. Catti, M. (2004) *J. Phys. Condens. Matter*, **16**, 3909.
58. Leoni, S. (2007) *Chem. Eur. J.*, **13**, 10022.
59. Tonkov, E.Yu (1992) *High Pressure Phase Transformations: A Handbook*, Gordon and Breach, London.
60. Sato-Sorensen, Y. (1983) *J. Geophys. Res.*, **88**, 3543.
61. Yagi, T., Suzuki, T., and Akimoto, S. (1983) *J. Phys. Chem. Solids*, **44**, 135.
62. Leger, J.M., Haines, J., Danneels, C., and de Oliveira, L.S. (1998) *J. Phys.: Condens. Matter*, **10**, 4201.
63. Laio, A. and Parrinello, M. (2002) *PNAS*, **99**, 12562.
64. Bärnighausen, H. (1980) *match*, **9**, 139.
65. Wondratschek, H. and Jeitschko, W. (1974) *Acta Crystallogr. A*, **30**, 431.
66. Broch, E., Oftedal, I., and Pabst, A. (1929) *Z. Phys. Chem. Abt. B*, **3**, 209.
67. Weir, C.E. and Piermarini, G.J. (1964) *J. Res. Natl. Bur. Stand., Sect. A*, **68**, 105.
68. Zahn, D., Hochrein, O., and Leoni, S. (2005) *Phys. Rev. B*, **72**, 094106.
69. Zahn, D. and Leoni, S. (2006) *J. Phys. Chem. B*, **110**, 10873.
70. Maier, J. (2004) *Nat. Mater.*, **4**, 805.
71. Maier, J. (2004) *Physical Chemistry of Ionic Materials: Ions and Electrons in Solids*, Wiley, Chichester.

72. Boulfelfel, S.E., Zahn, D., Hochrein, O., Grin, Yu., and Leoni, S. (2006) *Phys. Rev. B*, **74**, 094106.
73. Seifert, K.F. (1966) *Ber. Bunsenges. Phys. Chem.*, **70**, 1041.
74. Morris, E., Groy, T., and Leinenweber, K. (2001) *J. Phys. Chem. Solids*, **62**, 1117.
75. Hyde, B.G., O'Keeffe, M., Lyttle, W.M., and Brese, N.E. (1992) *Acta Chem. Scand.*, **46**, 216.
76. Gerward, L., Olsen, J.S., Steenstrup, S., Malinowski, M., Åsbrink, S., and Waskowskaet, A. (1992) *J. Appl. Crystallogr.*, **25**, 578.
77. Aroyo, M.I., Perez-Mato, J.M., Capillas, C., Kroumova, E., Ivantchev, S., Madariaga, G., Kirov, A., and Wondratschek, H. (2006) *Z. Kristallogr.*, **221**, 15.
78. Valle, M. (2005) *Z. Kristallogr.*, **220**, 585.
79. Maier, J. (1995) *Prog. Solid St. Chem.*, **23**, 171.
80. Chen, C.-C., Herhold, A.B., Johnson, C.S., and Alivisatos, A.P. (1997) *Science*, **276**, 398.
81. Zaziski, D., Prilliman, S., Scher, E.C., Casula, M., Wickham, J., Clark, S.M., and Alivisatos, P.A. (2004) *Nano Lett.*, **4**, 943.
82. Tolbert, S.H. and Alivisatos, A.P. (1994) *Science*, **265** 373.
83. Wickham, J.N., Herhold, A.B., and Alivisatos, A.P. (2000) *Phys. Rev. Lett.*, **84**, 923.
84. Mariano, A.N. and Warekois, E.P. (1963) *Science*, **142**, 672.
85. Sowa, H. (2007) *Solid State Sciences*, **7**, 1384.
86. Geddo Lehmann, A., Bionducci, M., and Buffa, F. (1998) *Phys. Rev B*, **58**, 5275.
87. Al'fer, S.A. and Skums, V.F. (2001) *Inorg. Mater.*, **37**, 1237.
88. Côtè, M., Zakharov, O., Rubio, A., and Cohen, M.L. (1997) *Phys. Rev. B*, **55**, 13025.
89. Rabani, E. (2002) *J. Chem. Phys.*, **116**, 258.
90. Leoni, S., Ramlau, R., Meier, K., Schmidt, M., and Schwarz, U. (2008) *PNAS*, **105**, 19612.
91. Schwarz, U., Takemura, K., Hanfland, M., and Syassen, K. (1998) *Phys. Rev. Lett.*, **81**, 2711.
92. Guloy, A.M., Ramlau, R., Tang, Z., Schnelle, W., Baitinger, M., and Grin, Yu. (2006) *Nature*, **443**, 320.
93. Raty, J.-Y., Schwegler, E., and Bonev, S.A. (2007) *Nature*, **449**, 448.
94. Hanfland, M., Schwarz, U., Syassen, K., and Takemura, K. (1999) *Phys. Rev. Lett.*, **82**, 1197.
95. Alivisatos, P. (2001) *Science*, **293**, 1803.
96. Behler, J., Martonak, R., Donadio, D., and Parrinello, M. (2008) *Phys. Rev. Lett.*, **100**, 185501.
97. Zapol, P., Pandey, R., and Gale, J. (1997) *J. Phys.: Condens. Matter*, **9**, 9517.
98. Chisholm, J.A., Lewis, D.W., and Bristowe, P.D. (1999) *J. Phys.: Condens. Matter*, **11**, L235.
99. Ponce, F.A. and Bour, D.P. (1997) *Nature*, **386**, 351.
100. Perlin, P., Jauberthie-Carillon, C., Itie, J.P., Miguel, A.S., Grzegory, I., and Polian, A. (1992) *Phys. Rev. B*, **45**, 83.
101. Chelikowsky, J.R. (1987) *Phys. Rev. B*, **35**, 1174.
102. Croll, J.A. (1967) *Phys. Rev.*, **157**, 623.
103. Tolbert, S.H. and Alivisatos, A.P. (1995) *J. Chem. Phys.*, **102**, 4642.
104. Sharma, S.M. and Gupta, Y.M. (1998) *Phys. Rev. B*, **58**, 5964.
105. Knudson, M.D., Gupta, Y.M., and Kunz, A.B. (1999) *Phys. Rev. B*, **59**, 11704.
106. Limpijumnong, S. and Lambrecht, W.R.L. (2001) *Phys. Rev. Lett.*, **86**, 91.
107. Marco Saitta, A. and Decremps, F. (2004) *Phys. Rev. B*, **70**, 035214.
108. Sowa, H. (2001) *Acta Crystallogr. A*, **57**, 176.
109. Tolbert, S.H. and Alivisatos, A.P. (1995) *J. Chem. Phys.*, **102**, 4642.
110. Donohue, P.C. (1970) *Inorganic Chem.*, **9**, 335.
111. Schwarz, U. and Syassen, K. (1992) *High Press. Res.*, **9**, 148.
112. Knudson, M.D. and Gupta, Y.M. (1999) *Phys. Rev. B*, **59**, 11704.
113. Liu, J.F., Yin, S., Wu, H.P., Zeng, Y.W., Hu, X.R., Wang, Y.W., Lv, G.L., and Jiang, J.Z. (2006) *J. Phy. Chem. B*, **110**, 21588.

114. Fierro, J.L.G. (2005) *Metal Oxides – Chemistry and Applications*, Taylor & Francis Group, LLC.
115. Özgür, Ü., Alivov, Ya.I., Liu, C., Teke, A., Reshchikov, M.A., Doğan, S., Avrutin, V., Cho, S.-J., and Morkoç, H (2005) *J. Appl. Phys.*, **98**, 041301.
116. Decremps, F., Pellicer-Porres, J., Datchi, F., Itié, J.P., Polian, A., Baudelet, F., and Jiang, J.Z. (2002) *Appl. Phys. Lett.*, **81**, 4820.
117. Bates, C.H., White, W.B., and Roy, R. (1962) *Science*, **137**, 993.
118. Decremps, F., Zhang, J., and Liebermann, R.C. (2000) *Europhys. Lett.*, **51**, 268.
119. Karzel, H., Potzel, W., and Kofferlein, M. (1996) *Phys. Rev. B*, **53**, 11425.
120. Desgreniers, S. (1998) *Phys. Rev. B*, **58**, 14102.
121. Gerward, L. and Olsen, J.S. (1995) *Synchroton Radiat.*, **2**, 233.
122. Liu, H., Ding, Y., Somayazulu, M., Qian, J., Shu, J., Häusermann, D., and Mao, H.K. (2005) *Phys. Rev. B*, **71**, 212103.
123. Jiang, J.Z., Olsen, J.S., Gerward, L., Frost, D., Rubie, D., and Peyronneau, J. (2000) *Europhy. Lett.*, **50**, 48.
124. Kulkarni, A.J., Sarasamak, K., Wang, J., Ke, F.J., Limpijumnong, S., and Zhou, M. (2008) *Mech. Res. Comm.*, **35**, 73.
125. Recio, J.M., Blanco, M.A., Lua na, V., Pandey, R., Gerward, L., and Staun Olsen, J. (1998) *Phys. Rev. B*, **58**, 8949.
126. Decremps, F., Zhang, J., Li, B., and Liebermann, R. (2001) *Phys. Rev. B*, **63**, 224105.
127. Manjón, F.J., Syassen, K., and Lauck, R. (2002) *High Press. Res.*, **22**, 299.
128. Mori, Y., Niiya, N., Ukegawa, K., Mizuno, T., Takarabe, K., and Ruoff, A.L. (2004) *phys. stat. sol.(b)*, **241**, 3198.
129. Liu, H., Tse, J.S., and Mao, H. (2006) *J. Appl. Phys.*, **100**, 093509.
130. Decremps, F., Datchi, F., Saitta, A.M., Polian, A., Pascarelli, S., Di Cicco, A., Itié, J.P., and Baudelet, F. (2003) *Phys. Rev. B*, **68**, 104101.
131. Mujica, A., Rubio, A., Mu noz, A., and Needs, R.J. (2003) *Rev. Mod. Phys.*, **75**, 863.
132. Sowa, H. and Ahsbahs, H. (2006) *J. Appl. Cryst.*, **39**, 169.
133. Decremps, F., Pellicer-Porres, J., Saitta, A.M., Chevrin, J.-C., and Polian, A. (2002) *Phys. Rev. B*, **65**, 092101.
134. Cai, J. and Chen, N. (2007) *J. Phys.: Condens. Matter.*, **19**, 266207.
135. Lewis, G.V. and Catlow, C.R.A. (1985) *J. Phys. C*, **18**, 1149.
136. Bayarjargal, L., Winkler, B., Haussühl, E., and Boehler, R. (2009) *Appl. Phys. Lett.*, **95**, 061907.

Appendix: First Blind Test of Inorganic Crystal Structure Prediction Methods

Artem R. Oganov, J. Christian Schön, Martin Jansen, Scott M. Woodley, William W. Tipton, and Richard G. Hennig

Inspired by the blind tests of (molecular) organic crystal structure prediction [1], we have decided to perform a blind test for non-molecular inorganic crystal structures. Since in organic structure prediction one deals with the packing of entire (and often rigid) molecules, many constraints are present and the actual number of degrees of freedom is often relatively small. Thus, the search space is not very large and the main challenge is to correctly rank the structures by energy. This requires accurate forcefields that incorporate diverse interactions (Coulomb, van der Waals, hydrogen-bridge, etc.); developing sufficiently accurate forcefields turns out to be a major, if not the main problem in the field of organic crystal structure prediction.

For inorganic non-molecular crystals, the number of degrees of freedom is $3N+3$ (where N is the number of atoms in the unit cell) – and this very often is a large number, reaching 100–300 for many practically important systems. Here, the combinatorial complexity of the problem is extremely high and increases exponentially with the number of atoms, and it is the NP-hard search problem that dominates the agenda. Such methods as random sampling (widely used in organic structure prediction [1]) are reliable for systems with up to $\sim$20–30 degrees of freedom, but turn out to be nearly useless for realistically large inorganic systems.

Thus, we have decided to test the ability of different search methods to find the lowest-energy structure, given the same forcefield. There is a fundamental difference from, and a major complementarity with the organic blind test [1] – we are testing the search algorithms, whereas previous blind tests only examined the success of energy ranking.

For our test, groups developing or practicing all major methods have been invited, and combining the results from the groups that accepted the challenge we were able to compare three approaches – evolutionary algorithm USPEX [2] (test done by A.R.O.), simulated annealing [3, 4] (test done by J.C.S, S.M.W. and M.J.) and random sampling [5, 6] (test done by W.W.T. and R.H.). For reviews of these methods, see [7–9] in this volume. The test cases were chosen to be at the current limit of the field – with up to 60 atoms in the unit cell and very complex chemical compositions that have not yet been synthesized. The latter requirement

Modern Methods of Crystal Structure Prediction. Edited by Artem R. Oganov
Copyright © 2011 WILEY-VCH Verlag GmbH & Co. KGaA, Weinheim
ISBN: 978-3-527-40939-6

excludes any experimental insight and makes the test truly "blind". We were interested to:

1) compare the success rate and efficiency of different approaches
2) find systematic trends in the performance of different methods
3) extract lessons from these results, in order to improve these methods and advance our field further.

As test cases, we took more-or-less randomly put together chemical compositions $BaMgAl_4Si_4O_{16}$ (in fixed cubic cells with one and two formula units – Test #1 and Test #2, respectively), $Mg_{10}Al_4Ge_2Si_8O_{36}$ (variable cell calculations – Test #3), and $Mg_{13}Al_8P_3$ (variable cell calculations – Test #4). Tests #1–#3 were performed using a rigid-ion model with full ionic charges, Buckingham short-range potentials with harmonic three-body O–Si–O terms and previously published [10–12] parameters. These calculations were done using the GULP code [13] coupled to structure prediction codes [2, 4, 7]. Test #4 employed *ab initio* total energy calculations, using the GGA functional [14] and all-electron PAW method [15], as implemented in the VASP code [16]. This last test turned out to be computationally quite expensive and was only attempted with USPEX and random sampling. All participating groups agreed to use the same computational parameters and same codes (GULP and VASP, respectively) for energy evaluation. Results from all three approaches are summarized in Table A.1.

To gain more insight into these results, we performed analysis based on fingerprint theory [17]. The fingerprint function, represented as a vector **F** in an abstract multidimensional space, leads to the following definitions of the cosine distance between structures 1 and 2:

$$D_{cosine} = \frac{1}{2}\left(1 - \frac{\mathbf{F}_1 * \mathbf{F}_2}{|\mathbf{F}_1||\mathbf{F}_2|}\right) \quad (A.1)$$

degree of order

$$\Pi = \frac{\Delta}{(V/N)^{1/3}}|\mathbf{F}|^2 \quad (A.2)$$

where Δ is the fingerprint discretization parameter (taken to be 0.10 Å in our calculations, but this parameter has little effect on the results) and V is the unit cell volume.

Another measure of disorder of a given structure is given by the structural quasientropy, which measures the average disparity between different positions occupied by atoms of the same element in the structure (we use a slightly updated definition, compared to the original one [17]):

$$S_{str} = -\sum_A \frac{N_A}{N_{cell}}\left\langle(1 - D_{A_iA_j})\ln(1 - D_{A_iA_j})\right\rangle \quad (A.3)$$

where the sum is over each atomic type A (N_A is the number of A-atoms in the unit cell), and $D_{A_iA_j}$ is the cosine distance (taking values between -1 and 1) between fingerprints of atoms in i-th and j-th positions. Degree of order Π and quasientropy

Table A.1 Performance of random sampling, simulated annealing and evolutionary algorithms for the four blind test cases[a,b].

	Random sampling	Simulated annealing	Evolutionary algorithm USPEX
Test #1, $BaMgAl_4Si_4O_{16}$ with fixed cubic cell (with forcefields)			
Number of runs (runs producing lowest E)	1 (1)	10 (1)	2 (2)
Minimum energy, eV	−876.94	**−877.99**	−877.71
Number of local optimizations before reaching the apparent ground state[c]	14794	7330	1465
Cosine distance from ground state	0.09	0.0	0.14
Degree of order	1.4	1.2	1.7
Quasientropy	0.143	0.167	0.128
Test #2, $Ba_2Mg_2Al_8Si_8O_{32}$ with fixed cubic cell (with forcefields)			
Number of runs (runs producing lowest E)	1 (1)	9 (1)	2 (1)
Minimum energy, eV	−1751.57	−1756.03	**−1757.14**
Number of local optimizations before reaching the apparent ground state[c]	14102	2435	3210
Cosine distance from ground state	0.06	0.09	0.0
Degree of order	0.69	0.71	0.87
Quasientropy	0.243	0.248	0.206
Test #3, $Mg_{10}Al_4Ge_2Si_8O_{36}$ with variable cell (with forcefields)			
Number of runs (runs producing lowest E)	1 (1)	9 (1)	1 (1)
Minimum energy, eV	−1943.46	−1949.10	**−1950.53**
Number of local optimizations before reaching the apparent ground state[c]	13029	685	4610
Cosine distance from ground state	0.05	0.05	0.0
Degree of order	0.91	0.95	1.15
Quasientropy	0.269	0.236	0.213
Test #4, $Mg_{13}Al_8P_3$ with variable cell (ab initio)			
Number of runs (runs producing lowest E)	1 (1)	–	1 (1)
Minimum energy, eV	−68.82	–	**−70.37**
Number of local optimizations before reaching the apparent ground state	978	–	4071

[a] Distances from the ground state, degrees of order and quasientropies are based on fingerprints computed with $R_{max} = 10$ Å, discretized with bin width $\Delta = 0.10$ Å, and Gaussian smoothing with $\sigma = 0.05$ Å.

[b] Random sampling runs for Tests #1–3 included 25,000 structures each, but in each test only about fourteen thousand structures could be successfully locally optimized, presumably due to singularities of the interatomic potential model used (this is, however, a very commonly used forcefield). Only correctly relaxed structures were considered here. The same singular behavior was seen in some of the evolutionary runs.

[c] During a simulated annealing run, the walker periodically performed five stochastic quenches (i.e. relaxations) from a stopping point. For simulated annealing, the entry for "number of local optimizations" denotes the number of relaxations – the number of stopping points was correspondingly 5 times smaller.

S_{str} are quite independent measures of structural disorder, and it is gratifying that their use in Table A.1 leads to similar conclusions.

The idea of this blind test was that each team puts a realistic amount of computational work in the test. Thus, only a limited number of runs were performed. Given that in each test each group found a different result, we cannot be fully sure that the ground state is not some structure that has escaped all of our runs. However, we are confident that even if this were the case, the ground state would be exceedingly hard to find with any of the presented methods. According to Table A.1, random sampling failed to produce the lowest-energy structure in all of the tests, in each test yielding much higher-energy structures than those produced using simulated annealing or evolutionary algorithm USPEX. This is in line with our earlier observations, where with USPEX we found energetically much superior structures, compared to random sampling, for high-pressure phases of silane SiH_4 ([18] compared to [19]), nitrogen ([20] compared to [21]), and stannane SnH_4 ([22] compared to [23]). Similarly, comparison studies of simulated annealing and random sampling have showed that the success rate of simulated annealing is considerably higher (simulated annealing was invented to go beyond random sampling minimization for glasses and other NP-hard systems!), even for very simple chemical systems [24].

In the three tests, where both USPEX and simulated annealing searches were performed, both methods were competitive in terms of produced energies (although at different levels of computational effort and with different levels of reproducibility – for each particular test, USPEX yielded much more similar/reproducible results, in terms of the lowest energy, than simulated annealing). USPEX won in two cases out of three (Tests #2 and #3), whereas simulated annealing won in Test #1. A summary of these tests is given in Table A.1 and Figures A.1–A.4.

For each run and optimization method, one might ask, how many more iterations should be completed after locating the "current" lowest energy structure? Running twice as many iterations may yield a better answer, but might also provide no improvement. By design, for the simulated annealing runs this is more or less a predetermined number, as the best structures are not expected (unless random search proves fortuitous) until lower temperatures are reached, which, by design, occurs in later iterations (e.g. for Test #1 and Test #2 the best structure was found at stopping points 1466 out of 1500 and 487 out of 500, respectively). In contrast, in Test #3, the best structure from the simulated annealing run was found during the early, high temperature, iterations, at stopping point 137 out of 500 (Figure A.3b). Clearly, just adding many more iterations would probably not improve the prediction for Test #3. Thus, in an improved second attempt (not performed in this study), one would want to employ a different temperature schedule, e.g. one where temperature would fall more slowly. Random sampling in principle always has the possibility of arriving at a better solution as the run length is increased; however, the probability of such an event becomes negligible for large systems, where the vast majority of generated random structures are physically identical [17]. For USPEX simulations, there are parameters that control, for example, the strength of the "natural selection" and magnitude of structural

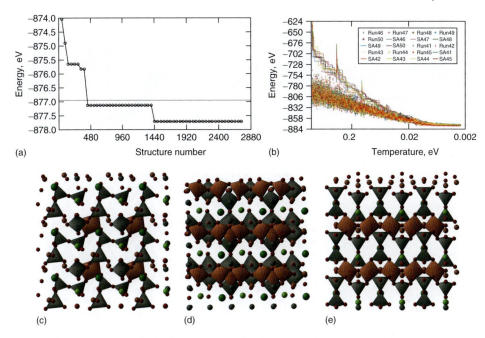

Figure A.1 Test #1 (BaMgAl$_4$Si$_4$O$_{16}$, with fixed cell): (a) Variation of the lowest energy during the evolutionary USPEX run, (b) Summary of simulated annealing runs (c–e) Lowest-energy structures obtained by random sampling, simulated annealing and USPEX, respectively. Shown are Al–O and Si–O polyhedra and Mg, Ba and O atoms are denoted as spheres of different sizes. Thin horizontal line in (a) shows the lowest energy found in 14794 random sampling attempts.

mutations – both factors affect the diversity of the population and thus its ability to evolve into better structures. In practice, we see that the diversity of the population is retained or even increases through the run, and therefore increasing the length of the run does offer a possibility of arriving at better solutions (this is clear, for example, from Figure A.3a, which implies that still better structures could be found in a longer run – but in this comparison of methods we restricted ourselves with runs of such length that can be done even at the *ab initio* level, for *practical* structure predictions). It should also be notes that rigorous comparison of efficiency of stochastic methods, such as simulated annealing or evolutionary algorithms, requires statistics over very large numbers of runs. Such statistical studies were not among the aims of the current test.

One should note as there are many more high energy structures (particularly when the number of variables increases), the random search approach is essentially like a quenching from various points along a path that could be created from the high temperature part of a simulated annealing run. Clearly, from the figures shown for the simulated annealing results of Tests #1 and #2 (Figures A.1a and A.1b, respectively), employing many more iterations for the random search approach is

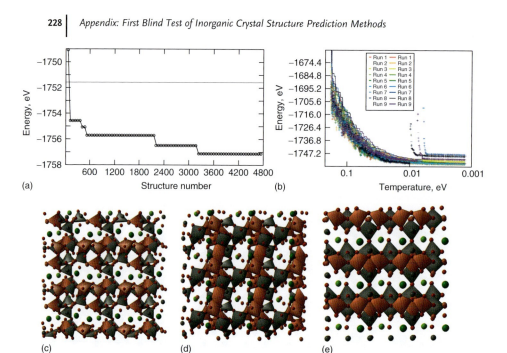

Figure A.2 Test #2 ($Ba_2Mg_2Al_8Si_8O_{32}$, with fixed cell): (a) Variation of the lowest energy during the evolutionary USPEX run, (b) Summary of simulated annealing runs, (c–e) Lowest-energy structures obtained by random sampling, simulated annealing and USPEX, respectively. Shown are Al–O and Si–O polyhedra and Mg, Ba and O atoms are denoted as spheres of different sizes. Thin horizontal line in (a) shows the lowest energy found in 14102 random sampling attempts.

unlikely to produce a similar "best" structure so far found by the other methods. Random search can also be viewed as the first (and energetically the poorest) generation of an evolutionary search, or an evolutionary search without selection and without any learning mechanism. Figures A.1a, A.2a, A.3a and A.4a show that in all cases, even short evolutionary runs outperform long and expensive random sampling runs. For Test #1, after 450 structure relaxations USPEX produces better structures than 14794 random sampling attempts. For Test #2, just 120 structure relaxations in USPEX produce better structures than 14102 relaxations in random sampling. For Test #3, within 60 structure relaxations we find better structures than after 13029 relaxations in random sampling. For Test #4, USPEX again wins – in 690 structure relaxations it found better structures than 978 random structure relaxations).

Structures produced by all methods were highly disordered (cf their low degrees of order and high quasientropies in Table A.1), which reflects the highly complex chemical compositions that probably do not correspond to chemical compounds that are stable against decomposition. In many ways, these optimization problems are reminiscent of trying to find the crystalline ground state of a system that exhibits a strong tendency to form amorphous structures (c.f. e.g. amorphous silicon boron

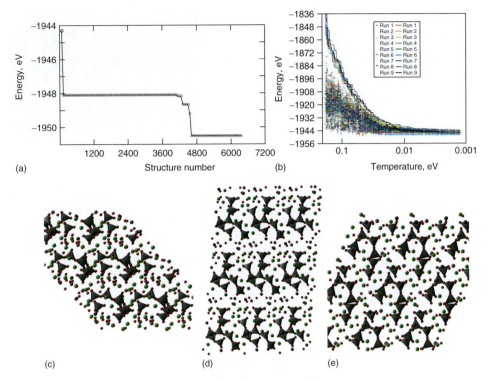

Figure A.3 Test #3 ($Mg_{10}Al_4Ge_2Si_8O_{36}$, variable cell): (a) Variation of the lowest energy during the evolutionary USPEX run, (b) Summary of simulated annealing runs, (c–e) Lowest-energy structures obtained by random sampling, simulated annealing and USPEX, respectively. Shown are Si–O polyhedra and Mg, Al, Ge and O atoms are denoted as spheres of different sizes. The lowest energy found in 13029 random sampling attempts is not shown in (a), because it is above the scale of the figure – i.e. is worse than the best structure even in the first generation. The reason is that in the USPEX simulation, the initial generation was produced using the random pseudosubcell technique [25] with splitting factors 6, 8, and 12.

nitride a-$Si_3B_3N_7$ [26] whose dynamics shows non-ergodic aging behaviour on typical simulation time scales without ever getting close to a crystalline ordered state [27]). This is also a reason why these tests were so challenging and computationally expensive, even with simple forcefields. In all cases, USPEX produced the most ordered solutions – which were also the best solutions in all tests, except Test #1. The best solutions found with simulated annealing and random sampling in all cases were equally disordered. As seen from cosine distances in Table A.1, best solutions found with simulated annealing and USPEX were maximally distant, with random sampling solutions either in between or at equal distance from both. The existence of such clear systematic trends in the behavior of these methods is unexpected and needs to be confirmed for other systems. If confirmed, it could offer new concepts and paths for improving these methods. This first

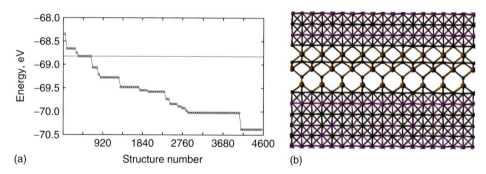

Figure A.4 Test #4 ($Mg_{13}Al_8P_3$, variable cell, *ab initio* calculation): (a) Variation of the lowest energy during the evolutionary USPEX run, (b) Lowest-energy structure obtained by USPEX. Thin horizontal line in (a) shows the lowest energy found in 978 random sampling attempts. The lowest-energy structure (b) indicates the tendency to phase separation and consists of layers of Mg-Al alloy with the close-packed fcc structure and layers of Mg,P-enriched open structure. Cases of phase separation are usually very challenging for structure prediction – (i) the initial random population usually consists of well-mixed atoms, and it takes many steps to rearrange them into a demixed state, (ii) such cases usually correspond to multi-funnel energy landscapes, which are intrinsically hard for global optimization, (iii) in this particular case the problem also includes finding the optimal Mg–Al ordering in the fcc-alloy layer.

blind test of inorganic crystal structure prediction shows the current state of the art, the advantages and disadvantages of different methods applied to a set of computationally very challenging problems. The authors believe that such tests need to be continued, retaining the spirit of openness and exchange of ideas (in the spirit of which, structural data and input/output files for the present test can be obtained on request from the authors). Such future tests will need to include more methods and address new aspects, such as statistics of performance of different methods (i.e. their efficiency and success rates).

References

1. Day, G.M., Motherwell, W.D.S., Ammon, H.L., Boerrigter, S.X.M., Della Valle, R.G. Venuti, E., Dzyabchenko, A., Dunitz, J.D., Schweizer, B., van Eijck, B.P., Erk, P., Facelli, J.C., Bazterra, V.E., Ferraro, M.B., Hofmann, D.W.M., Leusen, F.J.J., Liang, C., Pantelides, C.C., Karamertzanis, P.G., Price, S.L., Lewis, T.C., Nowell, H., Torrisi, A., Scheraga, H.A., Arnautova, Y.A., Schmidt, M.U., and Verwer, P. (2005) A third blind test of crystal structure prediction. *Acta Cryst.* **B61**, 511–527.

2. Oganov, A.R. and Glass, C.W. (2006) Crystal structure prediction using *ab initio* evolutionary techniques: principles and applications. *J. Chem. Phys.* **124**, art. 244704.

3. Pannetier, J., Bassasalsina, J., Rodriguez-Carva jal, J., and Caignaert, V. (1990) Prediction of crystal structures from crystal chemistry rules by simulated annealing. *Nature*, **346**, 343–345.

4. Schon, J.C. and Jansen, M. (1996) First step towards planning of syntheses in solid-state chemistry: determination of promising structure candidates by global

optimisation. *Angew. Chem.–Int. Ed.* **35**, 1287–1304.
5. Freeman, C.M., Newsam, J.M., Levine, S.M., and Catlow, C.R.A. (1993) Inorganic crystal structure prediction using simplified potentials and experimental unit cells – application to the polymorphs of titanium dioxide. *J. Mater. Chem.* **3**, 531–535.
6. Schmidt, M.U. and Englert, U. (1996) Prediction of crystal structures. *J. Chem. Soc. – Dalton Trans.* **10**, 2077–2082.
7. Tipton, W.W. and Hennig, R.G. (2010) Random search methods. This volume.
8. Schön, J.C. and Jansen, M. (2010) Predicting solid compounds using simulated annealing. This volume.
9. Lyakhov, A.O., Oganov, A.R., and Valle, M. (2010) Crystal structure prediction using evolutionary approach. This volume.
10. Lewis, G.V. and Catlow, C.R.A. (1985) Potential models for ionic oxides. *J. Phys. C.: Solid State Phys.* **18**, 1149–1161.
11. Gavezzotti, A. (1994) Are crystal structures predictable? *Acc. Chem. Res.* **27**, 309–314.
12. Sanders, M.J., Leslie, M., and Catlow, C.R.A. (1984) Interatomic potentials for SiO_2. *J. Chem. Soc., Chem. Commun.* **19**, 1271–1273.
13. Gale, J.D. (2005) GULP: Capabilities and prospects. *Z. Krist.* **220**, 552–554.
14. Perdew, J.P., Burke, K., and Ernzerhof, M. (1996) Generalized gradient approximation made simple. *Phys. Rev. Lett.* **77**, 3865–3868.
15. Kresse, G. and Joubert, D. (1999) From ultrasoft pseudopotentials to the projector augmented-wave method. *Phys. Rev.* **B59**, 1758–1775.
16. Kresse, G. and Furthmüller, J. (1996) Efficient iterative schemes for *ab initio* total-energy calculations using a plane wave basis set. *Phys. Rev.* **B54**, 11169–11186.
17. Oganov, A.R. and Valle, M. (2009) How to quantify energy landscapes of solids. *J. Chem. Phys.* **130**, 104504.
18. Martinez-Canales, M., Oganov, A.R., Lyakhov, A., Ma, Y., and Bergara, A. (2009) Novel structures of silane under pressure. *Phys. Rev. Lett.* **102**, 087005.
19. Pickard, C.J. and Needs, R.J. (2006) High-pressure phases of silane. *Phys. Rev. Lett.* **97**, art. 045504.
20. Ma, Y., Oganov, A.R., Xie, Y., Li, Z., and Kotakoski, J. (2009) Novel high pressure structures of polymeric nitrogen. *Phys. Rev. Lett.* **102**, 065501.
21. Pickard, C.J. and Needs, R.J. (2009) High-pressure phases of nitrogen. *Phys. Rev. Lett.* **102**, 125702.
22. Gao, G., Oganov, A.R., Li, Z., Li, P., Cui, T., Bergara, A., Ma, Y., Iitaka, T., and Zou, G. (2010) Crystal structures and superconductivity of stannane under high pressure. *Proc. Natl. Acad. Sci.* **107**, 1317–1320.
23. Pickard, C.J. and Needs, R.J. (2009) Structures at high pressure from random searching. *Phys. Status Solidi* **246**, 536–540.
24. Schön, J.C. and Jansen, M. (1994) Determination of candidate structures for Lennard-Jones crystals through cell optimisation. *Ber. Bunsenges. Phys. Chem.* **98**, 1541–1544.
25. Lyakhov, A.O., Oganov, A.R., and Valle, M. (2010) How to predict large and complex crystal structures. *Comp. Phys. Comm.* **181**, 1623–1632.
26. Hannemann, A., Schön, J.C., Jansen, M., Putz, H., Lengauer, T. (2004) Modeling amorphous $Si_3B_3N_7$: Structure and elastic properties. *Phys. Rev.* **B70**, 144201.
27. Hannemann, A., Schön, J.C., Jansen, M., and Sibani, P. (2005) Nonequilibrium dynamics in amorphous $Si_3B_3N_7$. *J. Phys. Chem.* **B109**, 11770–11776.

Color Plates

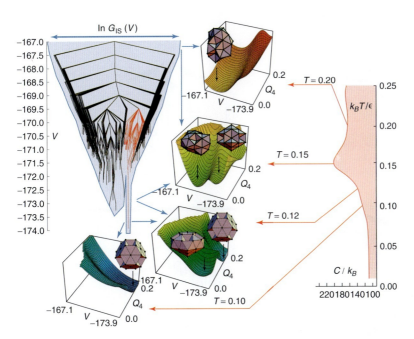

Figure 2.3 Disconnectivity graph, heat capacity, and free energy surfaces for the double funnel LJ_{38} cluster [42]. Here the width of the disconnectivity graph has been scaled according to the natural logarithm of the total number of minima with potential energy less than V, $\ln G_{IS}(V)$. A representative structure is illustrated above each free energy minimum. (This figure also appears on page 35.)

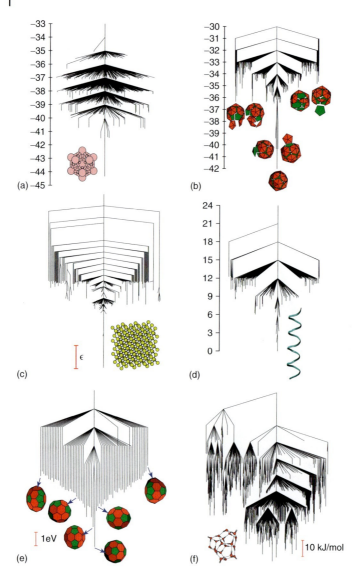

Figure 2.2 "Palm tree" (a)–(d) "willow tree" (e), and "banyan tree" (f) disconnectivity graphs for the following systems: (a) a cluster of 13 atoms bound by the Lennard–Jones (LJ) potential [61], LJ_{13}, including 1467 distinct local minima [62]. The energy is in units of the pair well depth. (b) An icosahedral shell composed of 12 pentagonal pyramids [41]. (c) A bulk representation of silicon using the Stillinger–Weber potential [63] and a supercell containing 216 atoms [64]. ϵ is the pair well depth. (d) The polyalanine peptide ala_{16} represented by the AMBER95 potential [65] and a distance-dependent dielectric [66]. The energy is in kcal/mol relative to the global minimum. (e) C_{60} using a density functional theory treatment of the electronic structure [67]. (f) An $(H_2O)_{20}$ cluster bound by the TIP4P [68] potential [49]. (This figure also appears on page 34.)

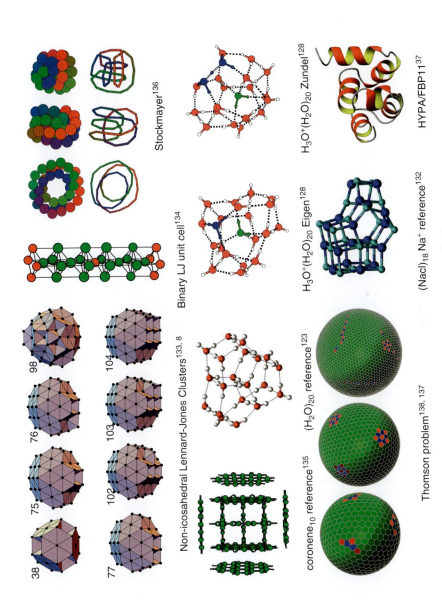

Figure 2.7 A selection of global minima from the Cambridge Cluster database at http://www.wales.ch.cam.ac.uk/CCD.html. (This figure also appears on page 40.)

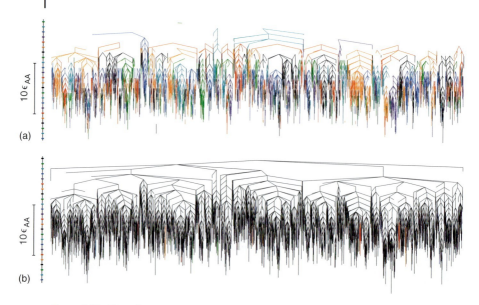

Figure 2.11 Two alternative representations of a disconnectivity graph obtained for a BLJ system with 60 atoms in the supercell at a number density of $1.3\sigma_{AA}^{-3}$ and $k_B T/\epsilon_{AA} = 0.713$. In (a), transition states corresponding to cage-breaking rearrangements are removed, while in (b) all the other transition states are omitted [172]. The graphs are colored according to the energy at which connection from the rest of the graph is lost, with a key on the vertical axis. The structure that results in (a) shows that the system can only explore local regions of configuration space when cage-breaking rearrangements are forbidden, leading to a higher order organization of the landscape. (This figure also appears on page 45.)

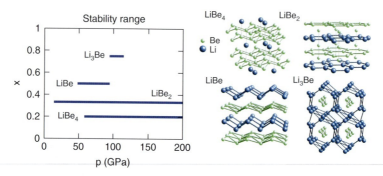

Figure 3.3 Li–Be compounds found at a high pressure using a random search method by Feng et al. [21]. (This figure also appears on page 62.)

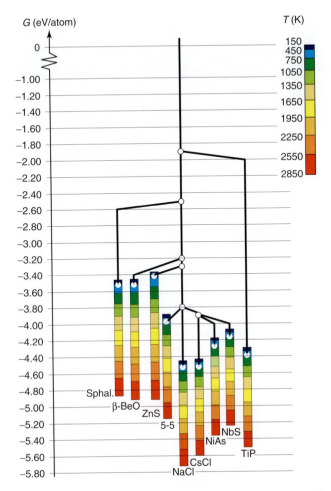

Figure 4.3 Free enthalpy landscape of SrO at $p = 0$ GPa for eight different temperatures ($T = 150$ K, ..., 2850 K) [42] computed using global landscape explorations followed by free energy calculations in the quasi-harmonic approximation on the empirical potential and *ab initio* level. The energetic contributions to the barriers stabilizing locally ergodic regions exhibiting different structure types are given by the energy difference between the minima (circles inside columns) and transition regions (white circles). Entropic barrier contributions (for a typical example see, e.g., [112]) are not shown to avoid overloading the figure. (This figure also appears on page 85.)

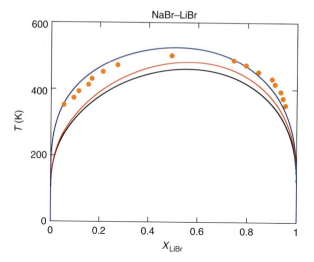

Figure 4.4 Low-temperature region of the phase diagram for the system NaBr–LiBr showing the miscibility gap in the system [28]. The gap was computed using global landscape explorations followed by the determination of free enthalpies employing both Hartree–Fock (black curve) and DFT–B3LYP (red curve) calculations. The blue curve is a fit to experimental data [191]; the yellow dots are experimental data points [192]. (This figure also appears on page 87.)

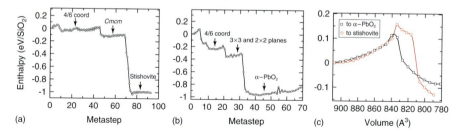

Figure 5.6 Evolution of the enthalpy during the transitions at 260 kbar and 600 K from cristobalite-XI to stishovite (a) and to α-PbO$_2$ (b). The activation barriers of the first step of the mechanisms (c) have been computed by optimizations of the atomic positions at the fixed h matrix. h values are determined by linear interpolation from cristobalite-XI to the first intermediates. After Ref. [41]. (This figure also appears on page 119.)

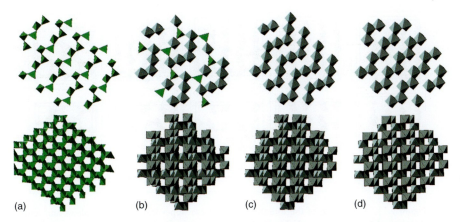

Figure 5.7 Section of a (11$\bar{2}$) plane (top row) and side view (bottom row) of the metastable structures encountered in the transition from cristobalite-XI (a) to α-PbO$_2$ (d). The transition occurs via the formation of a mixed tetrahedral and octahedral structure (b) and of a defective octahedral structure made of alternating 2×2 and 3×3 planes (c). After Ref. [41]. (This figure also appears on page 120.)

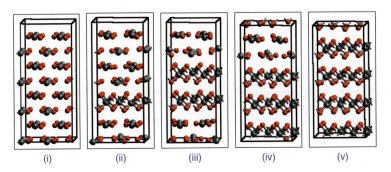

Figure 5.10 Structural evolution at steps 1, 4, 8, 86, and 89 in metadynamics simulations for a 32-molecule Phase II supercell of CO$_2$ at 60 GPa and 600 K. After Ref. [53]. (This figure also appears on page 122.)

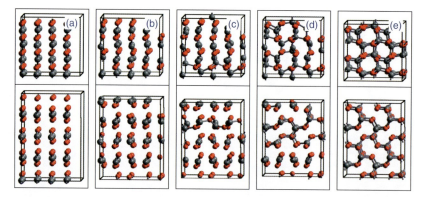

Figure 5.11 Illustration of the intermediate structures (metasteps 1, 14, 16, 17, 100) during the transformation of CO_2 from phase III (Cmca) to the α-cristobalite-like phase ($P4_12_12$) at 80 GPa and 300 K, where upper frames are top views and lower frames are side views. The density increases from 3.63 to 4.23 g/cm^3 from metastep 1 to metastep 100. After Ref. [53]. (This figure also appears on page 123.)

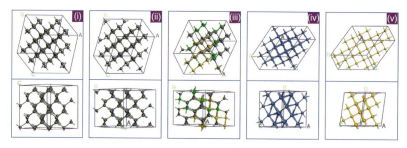

Figure 5.12 Metadynamics simulations of compressing cubic diamond at 2 TPa and 4000 K. Structural evolution of transformation from cubic diamond to SC1 (*Pm-3m*). After Ref. [60]. (This figure also appears on page 124.)

Color Plates | 241

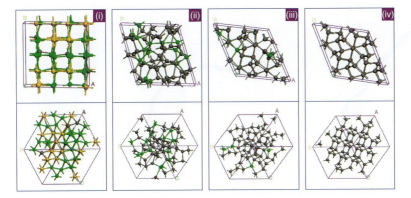

Figure 5.13 Structural evolution during metadynamics simulation yielding BC8 carbon (*Ia-3*) by decompressing SC1 (*Pm-3m*) carbon to 1 TPa at 5000 K. After Ref. [60]. (This figure also appears on page 124.)

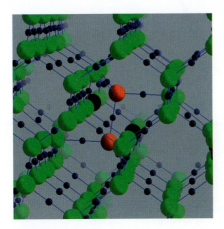

Figure 6.2 The green spheres show the silicon atoms in a perfect silicon crystal. The little blue dots represent the electrons which form chemical bonds. Each silicon atom has four bonds with its four nearest neighbors. Two silicon atoms were moved from the ¡?xmltex perfect?¿ crystal positions, pictured by the black spheres, to the new positions indicated by the two red spheres. The new system has again fourfold coordination and is a local minimum of the potential energy surface [3]. By repeated moves of this type, one can obtain crystalline structures with more and more point defects which finally become amorphous structures. (This figure also appears on page 133.)

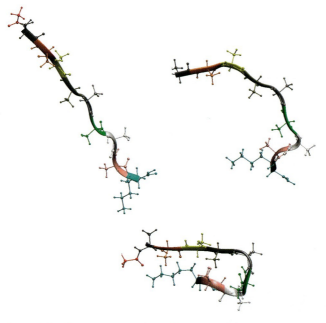

Figure 6.6 Three snapshots of an MD trajectory of polyalanin along a soft direction. (This figure also appears on page 138.)

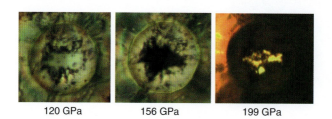

Figure 7.18 Photographs of sodium samples under pressure. At 120 GPa, the sample is metallic and highly reflective, at 156 GPa the reflectivity is very low, and at 199 GPa the sample is transparent. From Ma *et al.* [48]. (This figure also appears on page 169.)

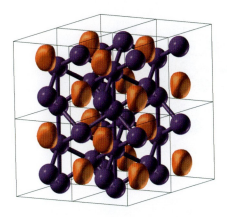

Figure 7.19 Crystal structure and electron localization function (isosurface contour 0.90) of the hP4 phase of sodium at 400 GPa. Interstitial electron localization is clearly seen. (This figure also appears on page 169.)

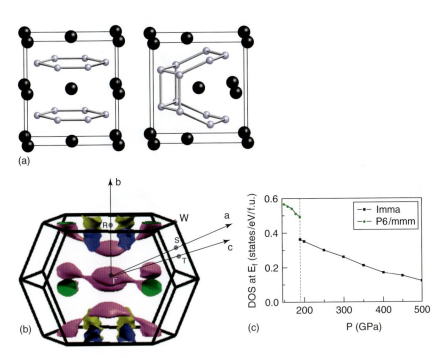

Figure 7.23 New phase of MgB_2: (a) its structure (right) and relationship to the structure known at 1 atm (left), (b) its Fermi surface, and (c) electronic density of states at the Fermi level as a function of pressure. From Ma et al. [49]. (This figure also appears on page 173.)

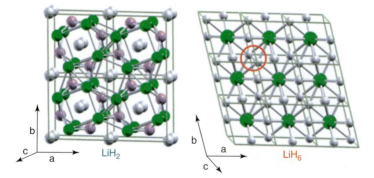

Figure 7.24 Structures of LiH_2 (a) and LiH_6 (b) predicted to be stable at pressures >100 GPa Li atoms are green, "lone" hydrogen atoms are pink, and those in the H_2 units are white. (from [63]). (This figure also appears on page 174.)

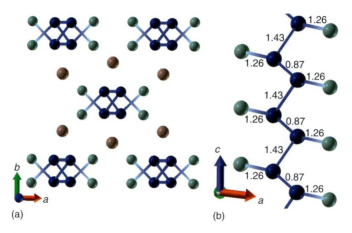

Figure 7.25 Crystal structure of the $C2/c$ phase of GeH_4. (a) Two types of hydrogen atoms are shown by different color (light–"lone" atoms, and dark–forming H_2 semimolecular units). Panel (b) shows the connectivity of hydrogens in the structure and distances (in Å). (This figure also appears on page 174.)

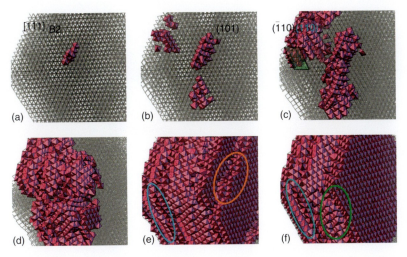

Figure 8.18 B2–B1 reconstruction in RbCl. Regions of B1 structure are marked with (pink) coordination polyhedra around Rb. (a) and (b) Initial nucleation events are spatially apart and uncorrelated. (b) and (c) Growth of larger regions of B1. Local formation of twin domains separated by trigonal prismatic polyhedra (green region), (c). (c)–(f) Further growth and fusion of B1 regions, under formation of smooth interfaces (blue ellipses), rough interfaces (green ellipses) and no interfaces (red ellipses). (This figure also appears on page 200.)

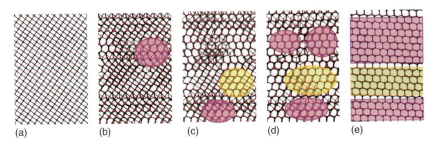

Figure 8.25 (a) Initial B1 configuration. (b)–(d) Nucleation of B4 (purple) and B3 (yellow) motifs, respectively. (e) Final lamellar arrangement. (This figure also appears on page 207.)

Index

a

A_xB_y binary Lennard–Jones system 165
ab initio total energy calculations 224, 227, 230
acceptance criterion 72, 75, 76
activation barrier 182
adaptive schedule 76
adjacency matrix 4, 5, 11
$Al_{12}C$ 172
AlB_2-type structure 163
alkali halides 186
AlN 210
ambient isotopy 2
augmentation 9
automorphism group 2, 4, 14–16

b

β-B_{106} 167
γ-B_{28} 165, 167
$BaCO_3$ 175
$BaMgAl_4Si_4O_{16}$ 224, 227
$Ba_2Mg_2Al_8Si_8O_{32}$ 228
banyan tree disconnectivity graph 33
barrier crossing 109, 110
barrier structure 67, 71, 79
basin of attraction 58
basin-hopping 29, 30, 36, 39, 41–43, 45, 156
bcc-based alloys 163
Bell-Evans-Polanyi principle 134–137, 139–141, 143
binary Lennard-Jones mixture 43
biological evolution 148
Boltzmann distribution 132, 142
Born Oppenheimer surface 131
boron 165–168
box matrix 111

c

C_{60} 33, 34
CaC_6 61
$CaCO_3$ 175
CaF_2 202
cage-breaking 44, 45
calcium 117, 125
Cambridge cluster database 40, 41
cancer growth phenomena 159
carbon 117, 120, 121, 123, 124
cartesian coordinate frame 162
cartesian distance 160
catchment basin 131, 133, 134, 137
CdSe 116, 117, 209
chemical intuition 62
chemical reaction 183
"chemical transmutation" operator 163
child structure 156
cluster structure prediction 162
CO_2 117, 120, 122, 123, 129
configuration space 68, 69, 73, 87, 98, 184
conformation family Monte Carlo algorithm 90
conjugate gradient 59
constrains for optimization, of energy landscapes 152
constraints 59
controlled disorder 70, 71, 98
convergence 60
coordination figure 3, 5, 9, 12, 15, 17, 19–24
coordination polymer 1, 12, 18, 19, 21, 23
cost function 68, 88, 94, 95
cotunnite 202
β-cristobalite structure 175
crystal structure prediction 150
crystallochemical database 11, 25
CsCl 186

d

damped molecular dynamics 59
decoration 9
defects 62
deluge algorithm 78
Demon algorithm 77
density functional theory 61, 142
disconnectivity graph 30–32, 34, 35, 39, 42, 43, 45
domain 194
double funnel 35
dual net 3–5, 17

e

earth's carbon cycle 175
edge net 4, 18, 25
edge-transitive net 17, 19
eigenvector-following 30a
"jellium" model 172
embedding 2, 3, 5, 10–16, 18, 22, 23
empirical potential 64
energy landscape 59, 67, 68, 70–73, 78, 80–82, 84, 86, 88, 90, 91, 93–98, 159, 171
entangled net 3
entropic stabilization 70
entropy 56
EPINET 7, 11, 13, 14
equilibration time 69
ergodicity 197
ergodicity search algorithm 68
escape time 69, 79
evolutionary algorithms 147–149, 152, 159
expansion 9

f

Fe–Mg system 163
fingerprint function 159–161, 224
fingerprint theory 161, 224
finite-time thermodynamics 76
first generation, initialization of 152–156
fitness function 149–150
fitness rank 157
fluorite 202
free energy 55
free energy disconnectivity graph 32
free energy landscapes 150, 164
Frenkel defect 204
frustration 33
funnel 153, 154, 158, 161

g

GaN 210
GeAs 213
GeH_4 173, 174
generalized barrier 71, 98
generalized gradient approximation (GGA) 164, 224
genetic algorithms 39
GeS 197
glass transition 44
global exploration 67, 68, 82, 84, 92
global minimum 56
global optimization 68, 72, 73, 75, 79, 81, 83, 84, 86, 88–92, 94–97
GMIN 37, 39
gold cluster 143
growth 181
GULP code 151, 224

h

$(H_2O)_{20}$ 33, 39
$H_6C_2O_6$ 175
$H_6Si_2O_7$ 175
halting criteria 158–159
hardness 198
heredity operator 149, 156, 161
heuristics 147
hierarchical approach 97, 98
high pressure 56, 61
hP4 phase of sodium 168, 169
HR-TEM 206
hydrogen-rich compounds 173
hysteresis 181, 186, 206, 207, 211, 214

i

icosahedral aluminum clusters 172
icosahedral B_{12} clusters 166
infinite graph 2
InN 211
inorganic crystal structure prediction, blind tests for 223–230
insulators by metal alloying 172–173
interface 191
interpenetrating net 3, 8
interstitial charge localization in sodium 170
intrinsic dimensionality, of the energy landscapes 151–153
inverse problem 55
iterative approach, for calculating lowest energy structure 226–228

k

KBr 198
KCl 197
KF 196
kinetic transition network 30

l

labeled quotient graph 4
landscape 29, 30, 32, 33, 36–39, 41, 42, 44, 45
latent heat 181
lattice mutation 155
lattice mutation operator 156
lattice vectors 152, 155
Lennard Jones cluster 135, 142
Lennard–Jones system 163
LiBe 62
LiH$_2$ 173, 174
LiH$_2$, LiH$_6$ and LiH$_8$ 173
LiH$_6$ 174
liquid 201
LJ$_{38}$ cluster 35, 39
LO–TO splitting 167
local free energy 69, 70, 98
local minimum 131, 133, 134, 136, 137, 141, 142
local optimization 56, 151–153
locally ergodic region 67–71, 75, 78, 79, 85, 86, 88, 90, 98
low-energy metastable states 157, 161, 171
lowest-energy structure 223, 226–227, 229–230

m

melting 141
metadynamics 108, 110–126
metal-organic framework MOF, 1
metastability 211
metastable carbon structures 171–172
metastable configuration 131
metastable phase 57
Metropolis algorithm 72, 133, 137, 138
Mg$_{10}$Al$_4$Ge$_2$Si$_8$O$_{36}$ 224, 229
Mg$_{13}$Al$_8$P$_3$ 224, 230
MgB$_2$ 172, 173
Mg–Fe system 166
MgSiO$_3$ 116–118, 159
minimum distance parameterization 59
Minkowski norm 160
molecular dynamics 182
molecular orientation, in organic crystals 162
Monte Carlo 183
Monte Carlo method 132, 133
morphology 197
moveclass 71, 72, 74–77, 93, 96–98
multi-overlap dynamics 77
multi-walker simulated annealing 77
multiple histogram approach 44
mutation operators 149, 155–157

n

NaBr 193
NaCl 183
NaCl-type arrangement 166
nanoclusters 62
natural tiling 3
net 1–20, 22–25
niching 161
non-crystallographic net 14–16
NP-hard problem 223
nucleation 108–110, 120, 126, 181

o

offspring 149
offspring structures 157–158
OPTIM 37
optimization problem 56
order parameter 110, 111, 114, 126, 184
organic crystal structure prediction 223
ζ-oxygen 170

p

palm tree disconnectivity graph 33
parallel tempering 77
parent structures 156
parents 149
Pareto optimization 94, 95
Parrinello-Rahman method 108, 110
path probability 185
PbCl$_2$ 202
periodic boundary conditions 58
periodic graph 1, 2
periodic nodal surface 187
perovskite 118
PES 29, 30, 44
phonon 210
polarizability 198
polycyclic aromatic hydrocarbons 41
polymerization 120, 121
polymorphism 202
population 149–150
post-perovskite 118
post-aragonite structure 175
potential energy surface 131, 133–136, 142, 183
pressure-induced phase transition 183
probability flow 79
protein folding 143
protein structure prediction 41
proteins 64

q

quasi-Newton method 59
quasientropy approach 224

r

radial distribution function (RDF)   160
Raman   215
random initialization   153
random sampling   152, 154, 223–226, 229
random search   56
rare event   182
RbCl   199
RCSR   3, 5, 6, 10–12, 14, 20
reaction coordinate   187
real-number crystal   155
real-number representation   150
reconstructive   181
regrouping   32
regular net   5
representation   150
reticular chemistry   2, 5, 9, 24
reverse Monte Carlo method   94
ring net   4, 18, 19, 25
rocksalt   211

s

search space   150
second order   181
secondary building unit (SBU)   24
selection   149
selection rules   161–163
self-dual net   5, 7, 17
semiconductor   205
shifting   184
shooting   184
SIESTA code   151
SiH_4 173
silane   61
silane SiH_4 226
silicate structure types   175
silicon   62
silicon cluster   137, 142, 143
similarity criteria   161
simulated annealing   68, 71–77, 79–84, 86, 88–92, 94–98, 223, 225–229
SnP   213
sodium   167–170
softness   198
solid state   181
solution space   59, 60
sphere packing   3, 5, 18, 19
$SrCO_3$ 175
stable stoichiometries of alloys   163
stannane SnH_4 226
steepest-descent path   30, 37
Stillinger-Weber silicon potential   43
stochastic exploration   77, 79, 81
stochastic quench   72, 74
stochastic search   57
stochastic selection   157
Stockmayer potential   41
stoichiometry   150, 163–164
stopping criteria   61
structural transformation   107–109, 114, 115, 120
structure determination   67, 79, 89, 91, 94, 95
structure prediction   68, 71, 75, 79–82, 84, 87, 90, 92, 94, 95, 97
structure representation   7, 8, 10, 11, 25, 58
subnet   2, 6, 14, 17
superconductivity   61
superhard graphite   171
supernet   2, 17
superposition approach   29
surviving structures   161
symmetry   194
Systre   10, 16

t

α-TlI   192
taboo search   77, 97
temperature cycling   76
temperature schedule   76, 96, 97
tetrahedral carbonate ions   175
Thomson problem   41
threshold algorithm   78, 79, 81, 84, 85
threshold value   152
tiling   3, 4, 7, 10
time scale problem   109, 110
TOPOS   10–12, 17
topotaxy   217
trajectory   184
trajectory space   184
transition mechanism   118, 126
transition path sampling   183
transition region   69, 71, 79, 85
transition state   29–32, 41, 44, 45
transitivity   4, 6, 16, 18
trapping   133, 134, 143
trial solution   58
Tsallis acceptance criterion   76

u

underlying net   8, 11, 16, 18–24
uninodal net   5, 15, 19, 20, 22
unit cell splitting   154–156
Universal Structure Predictor: Evolutionary Xtallography (USPEX)   150
USPEX   162–164, 168, 171, 172, 175
USPEX algorithm   223–226, 228, 229

v
variable cell 108, 125
variation operators 155–157
VASP code 151, 164, 224

w
willow tree disconnectivity graph 33
wurtzite 206

x
Xe–C system 171

z
zincblende 214
ZnO 205